Speech Enhancement

Speech Enhancement

A Signal Subspace Perspective

Jacob Benesty
Jesper Rindom Jensen
Mads Græsbøll Christensen
Jingdong Chen

ELSEVIER

AMSTERDAM • BOSTON • HEIDELBERG • LONDON
NEW YORK • OXFORD • PARIS • SAN DIEGO
SAN FRANCISCO • SINGAPORE • SYDNEY • TOKYO
Academic Press is an imprint of Elsevier

Academic Press is an imprint of Elsevier
The Boulevard, Langford Lane, Kidlington, Oxford, OX5 1GB
225 Wyman Street, Waltham, MA 02451, USA

First published 2014

British Library Cataloguing-in-Publication Data
A catalogue record for this book is available from the British Library

Library of Congress Cataloging-in-Publication Data
A catalog record for this book is available from the Library of Congress

ISBN: 978-0-12-800139-4

For information on all Academic Press publications
visit our website at **store.elsevier.com**

Printed and bound in United States of America

14 15 16 17 10 9 8 7 6 5 4 3 2 1

CONTENTS

CHAPTER 1

Introduction

In verbal communication, the presence of background noise, such as the sound of a passing car or an air vent, can impact the quality of the speech signal in a detrimental way, something that affects the listener and thus also the communication in several negative ways. Not only may the perceived quality of the speech be harmed, but also its intelligibility may be degraded. Even if only the perceived quality of the speech is affected, this may have a severe impact on the ability of the users to communicate, as exposure to noisy signals may cause listener fatigue. The presence of noise in signals is, though, not only a problem for humans. In speech processing systems, background noise causes additional problems, as such systems often comprise components that are designed under the assumption that only one, clean speech signal is present at any given time. This is, for example, the case for automatic speech recognizers and speech coders. This is typically done to simplify the design of these components, as the underlying statistical models then do not have to account for all possible noise types. Not only does this simplify the training of such models, it also, generally speaking, leads to faster algorithms; but it also renders these components vulnerable to noise.

As we have argued, the presence of background noise is problematic for humans and computers alike, and the problem of dealing with it, which is called speech enhancement or noise reduction, is an important and long-standing problem in signal processing (see, e.g., [1] and [2] for recent surveys), and the search for new and better methods continues today. Speech enhancement algorithms are important components in many systems, where speech plays a part, including telephony, hearing aids, voice over IP, and automatic speech recognizers. Speech enhancement is generally concerned with the problem of enhancing the quality of speech signals. This can, of course, mean many things, but it is often associated with the specific problem of reducing the impact of additive noise, which is also what we are concerned with in the present book. Additive noise occurs naturally in acoustic environments when multiple sources are present, and examples of common noise types are street, car, and babble.

Speech Enhancement. http://dx.doi.org/10.1016/B978-0-12-800139-4.00001-3

Moreover, it can also be caused by intrinsic noise in the sensor system, i.e., from the electrical components. To be more precise, the purpose of speech enhancement is to minimize the impact of the background noise while preserving the speech signal. Hence, there are two performance measures by which the efficiency of speech enhancement methods is compared: speech distortion and noise reduction [3]. These two measures are often conflicting, meaning that if we want to achieve the highest possible noise reduction, then we must accept speech distortion and, similarly, that if we cannot accept any speech distortion, then our ability to perform noise reduction will be hampered. An extreme example of this is the maximum signal-to-noise ratio filter [1] which achieves the highest possible noise reduction but at the cost of severe speech distortion.

The history of noise reduction can be traced back to the work of Wiener [4], i.e., to the very early days of signal processing. Due to the importance of the problem in particular in speech applications, many different solutions have been proposed over the years, and much time and effort is still devoted to the problem today. The problem is often broken into two sub-problems, namely the problem of finding a function to be applied to the observed signal so as to extract the desired signal, i.e., the speech signal, and the problem of finding the information that this function depends on. If we restrict ourselves to linear filters, then the first sub-problem is the problem of finding the optimal filter, i.e., a filter design problem. If the criterion for optimality is the mean-square error, then the so-called Wiener filter is the solution. This filter requires knowledge of the noise statistics (or the speech statistics), and the second sub-problem is then that of finding those statistics, often in the form of the noise correlation matrix or its power spectral density. In the past decade, most work seems to have focused on the second sub-problem, e.g., [5–9], under difficult conditions when the noise is nonstationary. This book is, however, concerned with the first sub-problem, which is determining the function that should be applied to the observed signal. This problem has, though, also seen some important new contributions regarding optimal filtering in the past few years, including [3, 10, 11].

In the literature, one can find many (seemingly) different attempts at solving the problem of speech enhancement, and at the time a new method is published, it is often not clear how exactly it relates to other, existing methods, often because it is either not clear exactly what problems are being solved, or that the problems are stated in different ways whose relation is difficult to ascertain. In fact, it appears that the retrospective process of relating methods may take decades, if it ever occurs. When listing existing classes of methods for speech enhancement, spectral subtraction, (optimal) linear filtering, statistical model-based approaches, and subspace methods are typically mentioned. Indeed, these are also the names of the chapters in the book [2]. The focus in the present book is on the class of methods generally known as optimal filtering, of which the classical Wiener filter is a special case. However, in this book, we will show how speech enhancement using the principles of subspace-based methods can be cast as an optimal filtering problem. As such, the present book unifies what has previously been considered two competing principles of speech enhancement in one framework. As a consequence, it is both possible to combine the benefits of the subspace methods and optimal filtering methods and to analyze and compare the performance of the various approaches analytically.

1.1 HISTORY AND APPLICATIONS OF SUBSPACE METHODS

The development of the subspace-based methods for speech enhancement took a quite different route than the more traditional speech enhancement methods based on the theory of stochastic processes (e.g., linear filtering methods), and it can, therefore, be quite difficult to understand similarities and differences between the methodologies. In that connection, the curious reader might wonder what exactly the distinguishing characteristics of subspace-based enhancement methods are. Subspace-based methods are a class of methods that take their starting point in linear algebra, i.e., they are based on the notions of subspaces and the properties of vectors and matrices. Simply put, they are based on the idea of decomposing the correlation matrix of the observed signal using an eigenvalue-type decomposition and then, from this, find a basis for the part of the space that contains the desired signal (called the signal subspace) and a basis for the part that contains only noise (called the noise subspace).

Subspace methods have a rich history in signal processing, not only for speech enhancement. In fact, much of the early work focused on problems such as parameter estimation, model order estimation, low-rank approximations, etc. Perhaps the earliest example of a subspace method for parameter estimation is Pisarenko's method [12] for sinusoidal parameter estimation. Later followed more, and probably the most famous, subspace methods for the same problem (although cast as the equivalent problem of determining spatial frequencies in arrays) such as the MUltiple SIgnal Classification (MUSIC) method [13, 14] (see also the later papers [15, 16]), of which Pisarenko's method is a special case, and the Estimation of Signal Parameters via Rotational Invariance Techniques (ESPRIT) method [17].

Since then, several variations, improvements, and generalizations have followed, including root MUSIC [18], modified MUSIC [19], min-norm [20], unitary ESPRIT [21], and (weighted) subspace fitting [22] (on this matter, see also the tutorial [23]). MUSIC exploits the orthogonality of the signal and noise subspaces while ESPRIT is based on exploiting the structure of the involved matrices, more specifically, their shift-invariance. In [24], it was shown how the model order can be determined statistically from the ratio between the arithmetic and geometric means of the eigenvalues in combination with model selection criteria (this was essentially based on the same derivations as [15]). Later, the ideas behind subspace methods lead to the more general ideas of reduced-rank signal processing [25] and low-rank adaptive filters [26]. More recently, it has been shown that the model order estimation problem can be solved not only based on the eigenvalues (as in [25]) but also by exploiting subspace orthogonality [27] and shift-invariance [28].

The roots of subspace-based enhancement methods can be traced back to [29], although that work appears at first glance to also deal with parameter estimation, focusing on frequency estimation using linear prediction. However, in the paper, enhanced signals are reconstructed via the singular value decomposition of the data matrix, and, hence, the first subspace-based signal enhancement method was born. Much of the early work focused on the simple white noise case [30–32] and later the more general case of colored noise was treated in detail in [33, 34] and later in [35]. In much of this work, the subspace-based enhancement problem is seen as a reduced-rank matrix approximation problem, wherein matrix decompositions are used to obtain a low-rank approximation of a data matrix. Since this approach has its roots in numerical linear algebra, the problem is then often seen as a deterministic one, where the realizations are being approximated, unlike statistical approaches and linear filtering based on stochastic processes. Although connections between matrix decompositions and estimation theory do exist [37] and a filtering interpretation of subspace methods was given in [36], the stated problems are quite different in nature.

At this point, it should be noted that within the general class of subspace methods, various decompositions have been used, like the eigenvalue decomposition, the singular value decomposition, and the Karhunen-Loève transform (e.g., [38, 39]), and some operate on correlation matrices, others on Toeplitz or Hankel data matrices. These are, however, mathematically equivalent, but their numerical properties and memory requirements may differ. Interestingly, triangular decompositions have also been considered more recently [40]. For more on the actual implementation of the various matrix decompositions and their properties, we refer the interested reader to [41]. As many real-time applications require not only fast computations but also sample-by-sample updates, fast methods for computing a basis for the signal or noise subspaces with time-recursive updates, so-called subspace trackers, have been developed [42–46]. Considering that fast and time-recursive implementations of linear filtering approaches are readily available, this is quite important in making the subspace-based methods practical.

1.2 SPEECH ENHANCEMENT FROM A SIGNAL SUBSPACE PERSPECTIVE

We will now go into a bit more detail about how subspace methods work. The simplest possible incarnation of a subspace method for enhancement is perhaps that of a projection of the observed signal onto a subspace known to contain the desired signal (and noise), i.e., the signal subspace. Then, any noise that may lie elsewhere, i.e., in the noise subspace, is removed and noise reduction is achieved without harming the speech signal. A number of questions then arise. First, how do we know that the desired signal, i.e., a speech signal, occupies only a subspace of the full space? Second, and if so, how do we identify this subspace?

To answer the first question, one can look to some commonly used models of speech signals. One such model is the harmonic model, which has a long and rich history in speech processing, specifically for modeling voiced speech. In that model, the speech signal is modeled as a sum of harmonically related sinusoids. For a specific number of harmonics, say C, such a model is well known to have a correlation matrix of rank $P = 2C$ (for the real case). Hence, for any $M \times M$ correlation matrix with $M > P$, the harmonics will lie in a subspace of dimension P. For signals that occupy the full space (whose correlation matrix have rank M), the justification for subspace methods can be found using the theory of low-rank approximation [25]. This theory states that the best rank r approximation of a $M \times M$ matrix (with $r < M$) is obtained by using the r largest singular values and the corresponding singular vectors, and the error (measured using the Frobenius norm) incurred by this is given simply by the remaining, small singular values. This applies, for example, when the number of harmonics of voiced speech exceeds the chosen dimension of the correlation matrix, and for autoregressive processes, which are often used as a model of unvoiced speech.

Returning now to the second question, i.e., how to identify the signal and noise subspaces, there are several ways in which this can be done. When the noise is white, the problem is particularly simple. In that case, the two subspaces can be identified from the eigenvalue decomposition of the observed signal correlation matrix, by simply sorting the eigenvectors according the magnitude of their eigenvalues. Then, the eigenvectors corresponding to the r largest eigenvalues span the same subspace as the desired signal, assuming that its correlation matrix is also rank r. Not only that, they form an orthonormal basis for that space (as the correlation matrices are symmetric by definition), and the eigenvectors corresponding to the remaining eigenvalues form an orthonormal basis for the noise subspace. It then also follows that the two sets of eigenvectors are orthogonal to each other. When the noise is colored, a pre-whitening step has to be included in the process, either as explicit pre-processing, e.g., in the form of filtering, or as part of the eigenvalue decomposition (see, e.g., [40]). More specifically, the appropriate decomposition is that of the generalized eigenvalue decomposition. Seen in a more general way, subspace-based speech enhancement can be seen as a modification of the eigenvalues via a diagonal so-called gain matrix. This way, subspace-based enhancement works by first transforming the signal vector, then applying the gain matrix, after which the signal vector is transformed back.

1.3 SCOPE AND ORGANIZATION OF THE WORK

The purpose of the present book is to unify the approaches of subspace-based enhancement and linear filtering, two approaches that have previously been considered separate classes of methods, and study the combined approach, both analytically as well as experimentally. We start out the book by introducing the general concept with diagonalization of the speech correlation matrix with the eigenvalue decomposition in Chapter 2. We introduce the basic signal model, along with all assumptions, and the basic problem formulation, and define some important quantities and the most important performance measures, namely input and output signal-to-noise ratios (SNRs), the noise reduction factor, the speech reduction factor, and speech distortion index (which are measures of speech distortion). We then proceed to derive several optimal rectangular filtering matrices based on the eigenvalue decomposition, namely the maximum SNR, Wiener, minimum variance distortionless response (MVDR), tradeoff, and linearly constrained minimum variance (LCMV) filters, something that we will continue to do for the various cases considered in the book. In Chapter 3, we then extend these principles to joint diagonalization of the speech and noise correlation matrices, i.e., using generalized eigenvalue decompositions, and analyze their performance. The problem of single-channel speech enhancement in the time domain is then addressed in Chapter 4 using the proposed framework. This is done in two cases: first for a rank-deficient speech correlation matrix, a case that applies, as previously explained, for voiced speech, and then for a full-rank speech correlation matrix. It is then demonstrated how to extend the principles from single-channel to multichannel speech enhancement in Chapter 5, still in the time domain. The generalization to multiple channels turns out to be somewhat complicated, but it leads to an approach that takes spatial information into account. In Chapter 6, the same problem is addressed, but this time in the frequency domain. This leads to a particularly simple solution for the binaural noise reduction problem. Chapter 7 explores a different problem yet, namely that of determining the speech (or signal) subspace using a promising, Bayesian approach based on the Stiefel manifold and the Bingham distribution. Finally, we study the performance of the various filters in simulations in Chapter 8. This is done using synthetic speech signals modeled using a set of harmonically related sinusoids and an autoregressive process, representing voiced and unvoiced speech, respectively. The former model results in speech correlation matrices that are rank deficient while the latter results in full-rank correlation matrices.

REFERENCES

[1] J. Benesty, J. Chen, Y. Huang, I. Cohen, Noise Reduction in Speech Processing, Springer-Verlag, Berlin, Germany, 2009.

[2] P. Loizou, Speech Enhancement: Theory and Practice, CRC Press, Boca Raton, FL, 2007.

[3] J. Benesty, J. Chen, Optimal Time-domain Noise Reduction Filters–A Theoretical Study, Springer Briefs in Electrical and Computer Engineering, Springer-Verlag, 2011.

[4] N. Wiener, Extrapolation, Interpolation, and Smoothing of Stationary Time Series: With Engineering Applications, M.I.T. Press, Boston, MA, 1949.

[5] S. Rangachari, P. Loizou, A noise estimation algorithm for highly nonstationary environments, Speech Commun. 28 (2006) 220–231.

[6] I. Cohen, Noise spectrum estimation in adverse environments: improved minima controlled recursive averaging, IEEE Trans. Speech Audio Process. 11 (2003) 466–475.

[7] T. Gerkmann, R.C. Hendriks, Unbiased MMSE-based noise power estimation with low complexity and low tracking delay, IEEE Trans. Speech Audio Process. 20 (2012) 1383–1393.

[8] R.C. Hendriks, R. Heusdens, J. Jensen, U. Kjems, Low complexity DFT-domain noise PSD tracking using high-resolution periodograms, EURASIP J. Adv. Signal Process. 2009 (2009) 15.

[9] J.R. Jensen, J. Benesty, M.G. Christensen, S.H. Jensen, Enhancement of single-channel periodic signals in the time-domain, IEEE Trans. Audio Speech Lang. Process. 20 (2012) 1948–1963.

[10] M.G. Christensen, A. Jakobsson, Optimal filter designs for separating and enhancing periodic signals, IEEE Trans. Signal Process. 58 (2010) 5969–5983.

[11] J.R. Jensen, J. Benesty, M.G. Christensen, S.H. Jensen, Non-causal time-domain filters for single-channel noise reduction, IEEE Trans. Audio Speech Lang. Process. 20 (2012) 1526–1541.

[12] V.F. Pisarenko, The retrieval of harmonics from a covariance function, Geophys. J. Roy. Astron. Soc. 33 (1973) 347–366.

[13] R.O. Schmidt, Multiple emitter location and signal parameter estimation, in: Proceedings of the RADC, Spectral Estimation Workshop, 1979, pp. 243–258.

[14] G. Bienvenu, Influence of the spatial coherence of the background noise on high resolution passive methods, Proc. IEEE ICASSP 4 (1979) 306–309.

[15] G. Bienvenu, L. Kopp, Optimality of high resolution array processing using the eigensystem approach, IEEE Trans. Acoust. Speech Signal Process. 31 (1983) 1235–1248.

[16] R.O. Schmidt, Multiple emitter location and signal parameter estimation, IEEE Trans. Antennas Propag. 34 (1986) 276–280.

[17] R. Roy, T. Kailath, ESPRIT – estimation of signal parameters via rotational invariance techniques, IEEE Trans. Acoust. Speech Signal Process. 37 (1989) 984–995.

[18] A. Barabell, Improving the resolution performance of eigenstructure-based direction-finding algorithms, Proc. IEEE ICASSP 8 (1983) 336–339.

[19] P. Stoica, K.C. sharman, Maximum likelihood methods for direction-of-arrival estimation, IEEE Trans. Acoust. Speech Signal Process. 38 (1990) 1132–1143.

[20] R. Kumaresan, D.W. Tufts, Estimating the angles of arrival of multiple plane waves, IEEE Trans. Aerosp. Electron. Syst. 19 (1983) 134–139.

[21] M. Haardt, J.A. Nossek, Unitary ESPRIT: how to obtain increased estimation accuracy with a reduced computational burden, IEEE Trans. Signal Process. 43 (1995) 1232–1242.

[22] M. Viberg, B. Ottersten, T. Kailath, Detection and estimation in sensor arrays using weighted subspace fitting, IEEE Trans. Signal Process. 39 (1991) 2436–2449.

[23] H. Krim, M. Viberg, Two decades of array signal processing research: the parametric approach, IEEE Signal Process. Mag. 13 (1996) 67–94.

[24] M. Wax, T. Kailath, Detection of the number of signals by information theoretic criterion, IEEE Trans. Acoust. Speech Signal Process. 33 (1985) 387–392.

[25] L.L. Scharf, The SVD and reduced rank signal processing, Signal Process. 25 (1991) 113–133.

[26] P. Strobach, Low-rank adaptive filters, IEEE Trans. Signal Process. 44 (1996) 2932–2947.

[27] M.G. Christensen, A. Jakobsson, S.H. Jensen, Sinusoidal order estimation using angles between subspaces, EURASIP J. Adv. Signal Process. 2009 (2009) 11.

[28] R. Badeau, B. David, G. Richard, A new perturbation analysis for signal enumeration in rotational invariance techniques, IEEE Trans. Signal Process. 54 (2006) 450–458.

[29] D.W. Tufts, R. Kumaresan, Singular value decomposition and improved frequency estimation using linear prediction, IEEE Trans. Acoust. Speech Signal Process. 30 (1982) 671–675.

[30] J.A. Cadzow, Signal enhancement-a composite property mapping algorithm, IEEE Trans. Acoust. Speech Signal Process. 30 (1988) 671–675.

[31] M. Dendrinos, S. Bakamidis, G. Carayannis, Speech enhancement from noise: a regenerative approach, Speech Commun. 10 (1991) 45–57.

[32] B. De Moor, The singular value decomposition and long and short spaces of noisy matrices, IEEE Trans. Signal Process. 41 (1993) 2826–2838.

[33] Y. Ephraim, H.L. Van Trees, A signal subspace approach for speech enhancement, IEEE Trans. Speech Audio Process. 3 (1995) 251–266.

[34] S.H. Jensen, P.C. Hansen, S.D. Hansen, J.A. Sørensen, Reduction of broad-band noise in speech by truncated QSVD, IEEE Trans. Speech Audio Process. 3 (1995) 439–448.

[35] P.S.K. Hansen, Signal subspace methods for speech enhancement, Ph.D. dissertation, Techn. Univ. Denmark, Lyngby, Denmark, 1997.

[36] P. Hansen, S. Jensen, FIR filter representations of reduced-rank noise reduction, IEEE Trans. Signal Process. 46 (1998) 1737–1741.

[37] T.W. Anderson, Asymptotic theory for principal component analysis, Ann. Math. Stat. 34 (1963) 122–148.

[38] U. Mittal, N. Phamdo, Signal/noise KLT based approach for enhancing speech degraded by colored noise, IEEE Trans. Speech Audio Process. 8 (2000) 159–167.

[39] A. Rezayee, S. Gazor, An adaptive KLT approach for speech enhancement, IEEE Trans. Speech Audio Process. 9 (2001) 87–95.

[40] P.C. Hansen, S.H. Jensen, Subspace-based noise reduction for speech signals via diagonal and triangular matrix decompositions: survey and analysis, EURASIP J. Adv. Signal Process. 2007 (2007) 24.

[41] G.H. Golub, C.F. Van Loan, Matrix Computations, third ed., The Johns Hopkins University Press, Baltimore, Maryland, 1996.

[42] B. Yang, Projection approximation subspace tracking, IEEE Trans. Signal Process. 41 (1995) 95–107.

[43] D. Rabideau, Fast, rank adaptive subspace tracking and applications, IEEE Trans. Signal Process. 44 (1996) 2229–2244.

[44] R. Badeau, G. Richard, B. David, Sliding window adaptive SVD algorithms, IEEE Trans. Signal Process. 52 (2004) 1–10.

[45] R. Badeau, B. David, G. Richard, Fast approximated power iteration subspace tracking, IEEE Trans. Signal Process. 53 (2005) 2931–2941.

[46] X. Doukopoulos, G. Moustakides, Fast and stable subspace tracking, IEEE Trans. Signal Process. 56 (2008) 1452–1465.

CHAPTER 2

General Concept with the Diagonalization of the Speech Correlation Matrix

In this chapter, we study the general speech enhancement problem by considering the diagonalization of the speech correlation matrix in the context of the classical linear filtering technique. The entire problem is formulated as a function of the speech subspace. We define the most fundamental performance measures in this scenario. We then derive the well-known conventional filtering matrices for noise reduction and show how the nullspace of the speech correlation matrix is exploited in some of these approaches such as the MVDR and LCMV filtering.

2.1 SIGNAL MODEL AND PROBLEM FORMULATION

We consider the general signal model:

$$\mathbf{y} = \mathbf{x} + \mathbf{v}, \tag{2.1}$$

where $\mathbf{y}$ is the observation or noisy signal vector of length M, $\mathbf{x}$ is the speech signal vector, and $\mathbf{v}$ is the noise signal vector. We assume that the components of the two vectors $\mathbf{x}$ and $\mathbf{v}$ are zero mean, stationary, and circular. We further assume that these two vectors are uncorrelated, i.e., $E\left(\mathbf{x}\mathbf{v}^H\right) = E\left(\mathbf{v}\mathbf{x}^H\right) = \mathbf{0}_{M\times M}$, where $E(\cdot)$ denotes mathematical expectation, the superscript H is the conjugate-transpose operator, and $\mathbf{0}_{M\times M}$ is a matrix of size $M \times M$ with all its elements equal to 0. In this context, the correlation matrix (of size $M \times M$) of the observations is

$$\begin{aligned}\mathbf{\Phi}_{\mathbf{y}} &= E\left(\mathbf{y}\mathbf{y}^H\right) \\ &= \mathbf{\Phi}_{\mathbf{x}} + \mathbf{\Phi}_{\mathbf{v}},\end{aligned} \tag{2.2}$$

where $\mathbf{\Phi}_{\mathbf{x}} = E\left(\mathbf{x}\mathbf{x}^H\right)$ and $\mathbf{\Phi}_{\mathbf{v}} = E\left(\mathbf{v}\mathbf{v}^H\right)$ are the correlation matrices of $\mathbf{x}$ and $\mathbf{v}$, respectively. In the rest of this chapter, we assume that the rank of the speech correlation matrix, $\mathbf{\Phi}_{\mathbf{x}}$, is equal to $P \leq M$ and the rank of the noise correlation matrix, $\mathbf{\Phi}_{\mathbf{v}}$, is equal to M. Then, the objective of speech enhancement (or noise reduction) is to estimate the desired signal vector, $\mathbf{x}$, or any known linear transformation of it from $\mathbf{y}$. This should

Speech Enhancement. http://dx.doi.org/10.1016/B978-0-12-800139-4.00002-5

be done in such a way that the noise is reduced as much as possible with no or little distortion of the speech signal vector.

Using the well-known eigenvalue decomposition [1], the speech correlation matrix can be diagonalized as

$$\mathbf{Q}_{\mathbf{x}}^H \mathbf{\Phi}_{\mathbf{x}} \mathbf{Q}_{\mathbf{x}} = \mathbf{\Lambda}_{\mathbf{x}}, \tag{2.3}$$

where

$$\mathbf{Q}_{\mathbf{x}} = \left[\mathbf{q}_{\mathbf{x},1} \; \mathbf{q}_{\mathbf{x},2} \cdots \mathbf{q}_{\mathbf{x},M} \right] \tag{2.4}$$

is a unitary matrix, i.e., $\mathbf{Q}_{\mathbf{x}}^H \mathbf{Q}_{\mathbf{x}} = \mathbf{Q}_{\mathbf{x}} \mathbf{Q}_{\mathbf{x}}^H = \mathbf{I}_M$, with $\mathbf{I}_M$ being the $M \times M$ identity matrix, and

$$\mathbf{\Lambda}_{\mathbf{x}} = \text{diag}\left(\lambda_{\mathbf{x},1}, \lambda_{\mathbf{x},2}, \ldots, \lambda_{\mathbf{x},M}\right) \tag{2.5}$$

is a diagonal matrix. The orthonormal vectors $\mathbf{q}_{\mathbf{x},1}, \mathbf{q}_{\mathbf{x},2}, \ldots, \mathbf{q}_{\mathbf{x},M}$ are the eigenvectors corresponding, respectively, to the eigenvalues $\lambda_{\mathbf{x},1}, \lambda_{\mathbf{x},2}, \ldots,$ $\lambda_{\mathbf{x},M}$ of the matrix $\mathbf{\Phi}_{\mathbf{x}}$, where $\lambda_{\mathbf{x},1} \geq \lambda_{\mathbf{x},2} \geq \cdots \geq \lambda_{\mathbf{x},P} > 0$ and $\lambda_{\mathbf{x},P+1} = \lambda_{\mathbf{x},P+2} = \cdots = \lambda_{\mathbf{x},M} = 0$. Let

$$\mathbf{Q}_{\mathbf{x}} = \left[\mathbf{T}_{\mathbf{x}} \; \mathbf{\Xi}_{\mathbf{x}} \right], \tag{2.6}$$

where the $M \times P$ matrix $\mathbf{T}_{\mathbf{x}}$ contains the eigenvectors corresponding to the nonzero eigenvalues of $\mathbf{\Phi}_{\mathbf{x}}$ and the $M \times (M - P)$ matrix $\mathbf{\Xi}_{\mathbf{x}}$ contains the eigenvectors corresponding to the null eigenvalues of $\mathbf{\Phi}_{\mathbf{x}}$. It can be verified that

$$\mathbf{I}_M = \mathbf{T}_{\mathbf{x}} \mathbf{T}_{\mathbf{x}}^H + \mathbf{\Xi}_{\mathbf{x}} \mathbf{\Xi}_{\mathbf{x}}^H. \tag{2.7}$$

Notice that $\mathbf{T}_{\mathbf{x}} \mathbf{T}_{\mathbf{x}}^H$ and $\mathbf{\Xi}_{\mathbf{x}} \mathbf{\Xi}_{\mathbf{x}}^H$ are two orthogonal projection matrices of rank P and $M - P$, respectively. Hence, $\mathbf{T}_{\mathbf{x}} \mathbf{T}_{\mathbf{x}}^H$ is the orthogonal projector onto the speech subspace (where all the energy of the speech signal is concentrated) or the range of $\mathbf{\Phi}_{\mathbf{x}}$, and $\mathbf{\Xi}_{\mathbf{x}} \mathbf{\Xi}_{\mathbf{x}}^H$ is the orthogonal projector onto the null subspace of $\mathbf{\Phi}_{\mathbf{x}}$. Using (2.7), we can write the speech vector as

$$\begin{aligned} \mathbf{x} &= \mathbf{Q}_{\mathbf{x}} \mathbf{Q}_{\mathbf{x}}^H \mathbf{x} \\ &= \mathbf{T}_{\mathbf{x}} \mathbf{T}_{\mathbf{x}}^H \mathbf{x} \\ &= \mathbf{T}_{\mathbf{x}} \widetilde{\mathbf{x}}, \end{aligned} \tag{2.8}$$

where

$$\widetilde{\mathbf{x}} = \mathbf{T}_{\mathbf{x}}^H \mathbf{x} \tag{2.9}$$

is the transformed desired signal vector of length P. Therefore, the signal model for noise reduction becomes

$$\mathbf{y} = \mathbf{T}_{\mathbf{x}} \widetilde{\mathbf{x}} + \mathbf{v}. \tag{2.10}$$

Fundamentally, from the M observations, we wish to estimate the P components of the transformed desired signal, i.e., $\widetilde{\mathbf{x}}$. Thanks to this transformation and the nullspace of $\mathbf{\Phi_x}$, we are able to reduce the dimension of the speech vector that we want to estimate. Indeed, there is no need to use the subspace $\mathbf{\Xi_x}$ since it contains no speech information. From (2.10), we give another form of the correlation matrix of $\mathbf{y}$:

$$\begin{aligned}\mathbf{\Phi_y} &= \mathbf{T_x}\mathbf{\Phi}_{\widetilde{\mathbf{x}}}\mathbf{T}_\mathbf{x}^H + \mathbf{\Phi_v} \\ &= \mathbf{T_x}\mathbf{\Lambda}_{\widetilde{\mathbf{x}}}\mathbf{T}_\mathbf{x}^H + \mathbf{\Phi_v},\end{aligned} \tag{2.11}$$

where

$$\begin{aligned}\mathbf{\Phi}_{\widetilde{\mathbf{x}}} &= E\left(\widetilde{\mathbf{x}}\widetilde{\mathbf{x}}^H\right) \\ &= \mathbf{T}_\mathbf{x}^H\mathbf{\Phi_x}\mathbf{T_x} \\ &= \mathbf{T}_\mathbf{x}^H\mathbf{Q_x}\mathbf{\Lambda_x}\mathbf{Q}_\mathbf{x}^H\mathbf{T_x} \\ &= \text{diag}\left(\lambda_{\mathbf{x},1}, \lambda_{\mathbf{x},2}, \ldots, \lambda_{\mathbf{x},P}\right) \\ &= \mathbf{\Lambda}_{\widetilde{\mathbf{x}}}\end{aligned} \tag{2.12}$$

and, obviously, $\mathbf{\Phi_x} = \mathbf{T_x}\mathbf{\Phi}_{\widetilde{\mathbf{x}}}\mathbf{T}_\mathbf{x}^H$.

2.2 LINEAR FILTERING WITH A RECTANGULAR MATRIX

From the general linear filtering approach [2], we can estimate the transformed desired signal vector, $\widetilde{\mathbf{x}}$, by applying a linear transformation to the observation signal vector, $\mathbf{y}$, i.e.,

$$\begin{aligned}\widetilde{\mathbf{z}} &= \widetilde{\mathbf{H}}\mathbf{y} \\ &= \widetilde{\mathbf{H}}\left(\mathbf{x} + \mathbf{v}\right) \\ &= \widetilde{\mathbf{x}}_{\text{fd}} + \widetilde{\mathbf{v}}_{\text{rn}},\end{aligned} \tag{2.13}$$

where $\widetilde{\mathbf{z}}$ is supposed to be the estimate of $\widetilde{\mathbf{x}}$,

$$\widetilde{\mathbf{H}} = \begin{bmatrix} \widetilde{\mathbf{h}}_1^H \\ \widetilde{\mathbf{h}}_2^H \\ \vdots \\ \widetilde{\mathbf{h}}_P^H \end{bmatrix} \tag{2.14}$$

is a rectangular filtering matrix of size $P \times M$,

$$\widetilde{\mathbf{h}}_p = \left[\widetilde{h}_{p,0}\ \widetilde{h}_{p,1} \cdots \widetilde{h}_{p,M-1}\right]^T, \quad p = 1, 2, \ldots, P \tag{2.15}$$

are complex-valued filters of length M, the superscript T is the transpose operator,

$$\begin{aligned}\widetilde{\mathbf{x}}_{\mathrm{fd}} &= \widetilde{\mathbf{H}}\mathbf{x} \\ &= \widetilde{\mathbf{H}}\mathbf{T}_{\mathbf{x}}\widetilde{\mathbf{x}}\end{aligned} \tag{2.16}$$

is the filtered (transformed) desired signal, and

$$\widetilde{\mathbf{v}}_{\mathrm{rn}} = \widetilde{\mathbf{H}}\mathbf{v} \tag{2.17}$$

is the residual noise. As a result, the estimate of $\mathbf{x}$ is supposed to be

$$\begin{aligned}\mathbf{z} &= \mathbf{T}_{\mathbf{x}}\widetilde{\mathbf{z}} \\ &= \mathbf{T}_{\mathbf{x}}\widetilde{\mathbf{H}}\mathbf{y} \\ &= \mathbf{H}\mathbf{y},\end{aligned} \tag{2.18}$$

where

$$\mathbf{H} = \mathbf{T}_{\mathbf{x}}\widetilde{\mathbf{H}} = \begin{bmatrix} \mathbf{h}_1^H \\ \mathbf{h}_2^H \\ \vdots \\ \mathbf{h}_M^H \end{bmatrix} \tag{2.19}$$

is the filtering matrix of size $M \times M$ that leads to the estimation of $\mathbf{x}$. The correlation matrix of $\widetilde{\mathbf{z}}$ is then

$$\begin{aligned}\mathbf{\Phi}_{\widetilde{\mathbf{z}}} &= E\left(\widetilde{\mathbf{z}}\widetilde{\mathbf{z}}^H\right) \\ &= \mathbf{\Phi}_{\widetilde{\mathbf{x}}_{\mathrm{fd}}} + \mathbf{\Phi}_{\widetilde{\mathbf{v}}_{\mathrm{rn}}},\end{aligned} \tag{2.20}$$

where

$$\begin{aligned}\mathbf{\Phi}_{\widetilde{\mathbf{x}}_{\mathrm{fd}}} &= \widetilde{\mathbf{H}}\mathbf{\Phi}_{\mathbf{x}}\widetilde{\mathbf{H}}^H \\ &= \widetilde{\mathbf{H}}\mathbf{T}_{\mathbf{x}}\mathbf{\Lambda}_{\widetilde{\mathbf{x}}}\mathbf{T}_{\mathbf{x}}^H\widetilde{\mathbf{H}}^H,\end{aligned} \tag{2.21}$$

$$\mathbf{\Phi}_{\widetilde{\mathbf{v}}_{\mathrm{rn}}} = \widetilde{\mathbf{H}}\mathbf{\Phi}_{\mathbf{v}}\widetilde{\mathbf{H}}^H. \tag{2.22}$$

We also observe that $\mathbf{\Phi}_{\mathbf{z}} = \mathbf{T}_{\mathbf{x}}\mathbf{\Phi}_{\widetilde{\mathbf{z}}}\mathbf{T}_{\mathbf{x}}^H$ and $\mathrm{tr}\left(\mathbf{\Phi}_{\mathbf{z}}\right) = \mathrm{tr}\left(\mathbf{\Phi}_{\widetilde{\mathbf{z}}}\right)$, where $\mathrm{tr}(\cdot)$ denotes the trace of a square matrix. The correlation matrix of $\widetilde{\mathbf{z}}$ or $\mathbf{z}$ is very helpful in the derivation of meaningful performance measures.

2.3 PERFORMANCE MEASURES

We are now ready to define the most important performance measures in the general context of speech enhancement described in Section 2.1.

We can divide these measures into two distinct but related categories. The first category evaluates the noise reduction performance while the second one evaluates speech distortion. We also discuss the very convenient mean-square error (MSE) criterion and show how it is related to the performance measures.

2.3.1 Noise Reduction

One of the most fundamental measures in all aspects of speech enhancement is the signal-to-noise ratio (SNR). The input SNR is a second-order measure, which quantifies the level of noise present relative to the level of the desired signal. It is defined as

$$\mathrm{iSNR} = \frac{\mathrm{tr}\left(\mathbf{\Phi}_{\mathbf{x}}\right)}{\mathrm{tr}\left(\mathbf{\Phi}_{\mathbf{v}}\right)}. \tag{2.23}$$

The output SNR, obtained from (2.20), helps quantify the SNR after filtering. It is given by

$$\begin{aligned}\mathrm{oSNR}\left(\widetilde{\mathbf{H}}\right) &= \frac{\mathrm{tr}\left(\mathbf{\Phi}_{\widetilde{\mathbf{x}}_{\mathrm{fd}}}\right)}{\mathrm{tr}\left(\mathbf{\Phi}_{\widetilde{\mathbf{v}}_{\mathrm{rn}}}\right)} \\ &= \frac{\mathrm{tr}\left(\widetilde{\mathbf{H}}\mathbf{\Phi}_{\mathbf{x}}\widetilde{\mathbf{H}}^{H}\right)}{\mathrm{tr}\left(\widetilde{\mathbf{H}}\mathbf{\Phi}_{\mathbf{v}}\widetilde{\mathbf{H}}^{H}\right)} \\ &= \frac{\sum_{p=1}^{P} \widetilde{\mathbf{h}}_p^H \mathbf{\Phi}_{\mathbf{x}} \widetilde{\mathbf{h}}_p}{\sum_{p=1}^{P} \widetilde{\mathbf{h}}_p^H \mathbf{\Phi}_{\mathbf{v}} \widetilde{\mathbf{h}}_p}.\end{aligned} \tag{2.24}$$

Then, the main objective of speech enhancement is to find an appropriate $\widetilde{\mathbf{H}}$ that makes the output SNR greater than the input SNR. Consequently, the quality of the noisy signal may be enhanced. It can be shown that [3]

$$\mathrm{oSNR}\left(\widetilde{\mathbf{H}}\right) \leq \max_{p} \frac{\widetilde{\mathbf{h}}_p^H \mathbf{\Phi}_{\mathbf{x}} \widetilde{\mathbf{h}}_p}{\widetilde{\mathbf{h}}_p^H \mathbf{\Phi}_{\mathbf{v}} \widetilde{\mathbf{h}}_p}, \tag{2.25}$$

which implies that

$$\mathrm{oSNR}\left(\widetilde{\mathbf{H}}\right) \leq \lambda_1, \tag{2.26}$$

where λ_1 is the maximum eigenvalue of the matrix $\mathbf{\Phi}_{\mathbf{v}}^{-1}\mathbf{\Phi}_{\mathbf{x}}$, whose value is always strictly positive. This shows how the output SNR is upper bounded. It is easy to check that

$$\begin{aligned}\text{oSNR}\left(\mathbf{H}\right) &= \frac{\text{tr}\left(\mathbf{H}\mathbf{\Phi}_{\mathbf{x}}\mathbf{H}^H\right)}{\text{tr}\left(\mathbf{H}\mathbf{\Phi}_{\mathbf{v}}\mathbf{H}^H\right)} \\ &= \text{oSNR}\left(\widetilde{\mathbf{H}}\right) \end{aligned} \tag{2.27}$$

and

$$\text{oSNR}\left(\mathbf{H}\right) \leq \max_{m} \frac{\mathbf{h}_m^H \mathbf{\Phi}_{\mathbf{x}} \mathbf{h}_m}{\mathbf{h}_m^H \mathbf{\Phi}_{\mathbf{v}} \mathbf{h}_m}. \tag{2.28}$$

Fundamentally, there is no difference between $\widetilde{\mathbf{H}}$ and $\mathbf{H}$. Both matrices lead to the same result as we should expect.

The noise reduction factor quantifies the amount of noise being rejected by $\widetilde{\mathbf{H}}$. This quantity is defined as the ratio of the power of the original noise over the power of the noise remaining after filtering, i.e.,

$$\begin{aligned}\xi_{\text{nr}}\left(\widetilde{\mathbf{H}}\right) &= \frac{\text{tr}\left(\mathbf{\Phi}_{\mathbf{v}}\right)}{\text{tr}\left(\widetilde{\mathbf{H}}\mathbf{\Phi}_{\mathbf{v}}\widetilde{\mathbf{H}}^H\right)} \\ &= \xi_{\text{nr}}\left(\mathbf{H}\right). \end{aligned} \tag{2.29}$$

Any good choice of $\widetilde{\mathbf{H}}$ should lead to $\xi_{\text{nr}}\left(\widetilde{\mathbf{H}}\right) \geq 1$.

2.3.2 Speech Distortion

The transformed desired speech signal can be distorted by the rectangular filtering matrix. Therefore, the speech reduction factor is defined as

$$\begin{aligned}\xi_{\text{sr}}\left(\widetilde{\mathbf{H}}\right) &= \frac{\text{tr}\left(\mathbf{\Phi}_{\mathbf{x}}\right)}{\text{tr}\left(\mathbf{\Phi}_{\widetilde{\mathbf{x}}_{\text{fd}}}\right)} \\ &= \frac{\text{tr}\left(\mathbf{\Lambda}_{\widetilde{\mathbf{x}}}\right)}{\text{tr}\left(\widetilde{\mathbf{H}}\mathbf{T}_{\mathbf{x}}\mathbf{\Lambda}_{\widetilde{\mathbf{x}}}\mathbf{T}_{\mathbf{x}}^H\widetilde{\mathbf{H}}^H\right)} \\ &= \xi_{\text{sr}}\left(\mathbf{H}\right). \end{aligned} \tag{2.30}$$

From (2.16), a rectangular filtering matrix that does not affect the transformed desired signal requires the constraint:

$$\widetilde{\mathbf{H}}\mathbf{T}_{\mathbf{x}} = \mathbf{I}_P, \tag{2.31}$$

where $\mathbf{I}_P$ is the $P \times P$ identity matrix. Hence, $\xi_{\text{sr}}\left(\widetilde{\mathbf{H}}\right) = 1$ in the absence of distortion and $\xi_{\text{sr}}\left(\widetilde{\mathbf{H}}\right) > 1$ in the presence of distortion. Taking the minimum ℓ_2-norm solution of (2.31), we get

$$\begin{aligned}\widetilde{\mathbf{H}} &= \left(\mathbf{T}_{\mathbf{x}}^H \mathbf{T}_{\mathbf{x}}\right)^{-1} \mathbf{T}_{\mathbf{x}}^H \\ &= \mathbf{T}_{\mathbf{x}}^H. \end{aligned} \tag{2.32}$$

This solution corresponds to the MVDR filter with white noise (see Section 2.4.3).

By making the appropriate substitutions, one can derive the relationship among the measures defined so far, i.e.,

$$\frac{\text{oSNR}\left(\widetilde{\mathbf{H}}\right)}{\text{iSNR}} = \frac{\xi_{\text{nr}}\left(\widetilde{\mathbf{H}}\right)}{\xi_{\text{sr}}\left(\widetilde{\mathbf{H}}\right)}. \tag{2.33}$$

When no distortion occurs, the gain in SNR coincides with the noise reduction factor.

Another way to measure the distortion of the transformed desired speech signal due to the filtering operation is via the speech distortion index defined as [4]

$$\begin{aligned}\upsilon_{\text{sd}}\left(\widetilde{\mathbf{H}}\right) &= \frac{E\left[\left(\widetilde{\mathbf{x}}_{\text{fd}} - \widetilde{\mathbf{x}}\right)^H \left(\widetilde{\mathbf{x}}_{\text{fd}} - \widetilde{\mathbf{x}}\right)\right]}{\text{tr}\left(\mathbf{\Lambda}_{\widetilde{\mathbf{x}}}\right)} \\ &= \frac{\text{tr}\left[\left(\widetilde{\mathbf{H}}\mathbf{T}_{\mathbf{x}} - \mathbf{I}_P\right) \mathbf{\Lambda}_{\widetilde{\mathbf{x}}} \left(\widetilde{\mathbf{H}}\mathbf{T}_{\mathbf{x}} - \mathbf{I}_P\right)^H\right]}{\text{tr}\left(\mathbf{\Lambda}_{\widetilde{\mathbf{x}}}\right)} \\ &= \upsilon_{\text{sd}}\left(\mathbf{H}\right). \end{aligned} \tag{2.34}$$

The speech distortion index is always greater than or equal to 0 and should be upper bounded by 1 for optimal rectangular filtering matrices; so the higher is the value of $\upsilon_{\text{sd}}\left(\widetilde{\mathbf{H}}\right)$, the more the transformed desired signal is distorted.

2.3.3 MSE Criterion

Since the transformed desired signal is a vector of length P, so is the error signal. We define the error signal vector between the estimated and desired signals as

$$\begin{aligned}\widetilde{\mathbf{e}} &= \widetilde{\mathbf{z}} - \widetilde{\mathbf{x}} \\ &= \widetilde{\mathbf{H}}\mathbf{y} - \widetilde{\mathbf{x}}, \end{aligned} \tag{2.35}$$

which can also be expressed as the sum of two uncorrelated error signal vectors:

$$\widetilde{\mathbf{e}} = \widetilde{\mathbf{e}}_{\text{ds}} + \widetilde{\mathbf{e}}_{\text{rs}}, \tag{2.36}$$

where

$$\begin{aligned}\widetilde{\mathbf{e}}_{\text{ds}} &= \widetilde{\mathbf{x}}_{\text{fd}} - \widetilde{\mathbf{x}} \\ &= \left(\widetilde{\mathbf{H}}\mathbf{T}_{\mathbf{x}} - \mathbf{I}_P\right)\widetilde{\mathbf{x}}\end{aligned} \tag{2.37}$$

is the signal distortion due to the rectangular filtering matrix and

$$\begin{aligned}\widetilde{\mathbf{e}}_{\text{rs}} &= \widetilde{\mathbf{v}}_{\text{rn}} \\ &= \widetilde{\mathbf{H}}\mathbf{v}\end{aligned} \tag{2.38}$$

represents the residual noise. Therefore, the mean-square error (MSE) criterion is

$$\begin{aligned}J\left(\widetilde{\mathbf{H}}\right) &= \text{tr}\left[E\left(\widetilde{\mathbf{e}}\,\widetilde{\mathbf{e}}^H\right)\right] \\ &= \text{tr}\left(\mathbf{\Lambda}_{\widetilde{\mathbf{x}}}\right) + \text{tr}\left(\widetilde{\mathbf{H}}\mathbf{\Phi}_{\mathbf{y}}\widetilde{\mathbf{H}}^H\right) - \text{tr}\left(\widetilde{\mathbf{H}}\mathbf{T}_{\mathbf{x}}\mathbf{\Lambda}_{\widetilde{\mathbf{x}}}\right) - \text{tr}\left(\mathbf{\Lambda}_{\widetilde{\mathbf{x}}}\mathbf{T}_{\mathbf{x}}^H\widetilde{\mathbf{H}}^H\right).\end{aligned} \tag{2.39}$$

Using the fact that $E\left(\widetilde{\mathbf{e}}_{\text{ds}}\widetilde{\mathbf{e}}_{\text{rs}}^H\right) = \mathbf{0}_{P\times P}$, $J\left(\widetilde{\mathbf{H}}\right)$ can be expressed as the sum of two other MSEs, i.e.,

$$\begin{aligned}J\left(\widetilde{\mathbf{H}}\right) &= \text{tr}\left[E\left(\widetilde{\mathbf{e}}_{\text{ds}}\widetilde{\mathbf{e}}_{\text{ds}}^H\right)\right] + \text{tr}\left[E\left(\widetilde{\mathbf{e}}_{\text{rs}}\widetilde{\mathbf{e}}_{\text{rs}}^H\right)\right] \\ &= J_{\text{ds}}\left(\widetilde{\mathbf{H}}\right) + J_{\text{rs}}\left(\widetilde{\mathbf{H}}\right),\end{aligned} \tag{2.40}$$

where

$$\begin{aligned}J_{\text{ds}}\left(\widetilde{\mathbf{H}}\right) &= \text{tr}\left[\left(\widetilde{\mathbf{H}}\mathbf{T}_{\mathbf{x}} - \mathbf{I}_P\right)\mathbf{\Lambda}_{\widetilde{\mathbf{x}}}\left(\widetilde{\mathbf{H}}\mathbf{T}_{\mathbf{x}} - \mathbf{I}_P\right)^H\right] \\ &= \text{tr}\left(\mathbf{\Lambda}_{\widetilde{\mathbf{x}}}\right)\upsilon_{\text{sd}}\left(\widetilde{\mathbf{H}}\right)\end{aligned} \tag{2.41}$$

and

$$\begin{aligned}J_{\text{rs}}\left(\widetilde{\mathbf{H}}\right) &= \text{tr}\left(\widetilde{\mathbf{H}}\mathbf{\Phi}_{\mathbf{v}}\widetilde{\mathbf{H}}^H\right) \\ &= \frac{\text{tr}\left(\mathbf{\Phi}_{\mathbf{v}}\right)}{\xi_{\text{nr}}\left(\widetilde{\mathbf{H}}\right)}.\end{aligned} \tag{2.42}$$

We deduce that

$$\begin{aligned}\frac{J_{\text{ds}}\left(\widetilde{\mathbf{H}}\right)}{J_{\text{rs}}\left(\widetilde{\mathbf{H}}\right)} &= \text{iSNR} \times \xi_{\text{nr}}\left(\widetilde{\mathbf{H}}\right) \times \upsilon_{\text{sd}}\left(\widetilde{\mathbf{H}}\right) \\ &= \text{oSNR}\left(\widetilde{\mathbf{H}}\right) \times \xi_{\text{sr}}\left(\widetilde{\mathbf{H}}\right) \times \upsilon_{\text{sd}}\left(\widetilde{\mathbf{H}}\right).\end{aligned} \tag{2.43}$$

We observe how the MSEs are related to the different performance measures.

2.4 OPTIMAL RECTANGULAR FILTERING MATRICES

In this section, we derive the most important rectangular filtering matrices that can help reduce the level of the noise. We will see how these optimal matrices depend explicitly on the speech subspace and, in some cases, how the nullspace of $\mathbf{\Phi}_{\mathbf{x}}$ is exploited.

2.4.1 Maximum SNR

From Section 2.3.1, we know that the output SNR is upper bounded by λ_1, which we can consider as the maximum possible output SNR. Then, it is easy to verify that with

$$\widetilde{\mathbf{H}}_{\max} = \begin{bmatrix} \varsigma_1 \mathbf{b}_1^H \\ \varsigma_2 \mathbf{b}_1^H \\ \vdots \\ \varsigma_P \mathbf{b}_1^H \end{bmatrix}, \tag{2.44}$$

where ς_p, $p = 1, 2, \ldots, P$ are arbitrary complex numbers with at least one of them different from 0 and $\mathbf{b}_1$ is the eigenvector of the matrix $\mathbf{\Phi}_{\mathbf{v}}^{-1}\mathbf{\Phi}_{\mathbf{x}}$ corresponding to λ_1, we have

$$\text{oSNR}\left(\widetilde{\mathbf{H}}_{\max}\right) = \lambda_1. \tag{2.45}$$

As a consequence, $\widetilde{\mathbf{H}}_{\max}$ can be considered as the maximum SNR filtering matrix. Clearly,

$$\text{oSNR}\left(\widetilde{\mathbf{H}}_{\max}\right) \geq \text{iSNR} \tag{2.46}$$

and

$$0 \leq \text{oSNR}\left(\widetilde{\mathbf{H}}\right) \leq \text{oSNR}\left(\widetilde{\mathbf{H}}_{\max}\right), \forall \widetilde{\mathbf{H}}. \tag{2.47}$$

The choice of the values of ς_p, $p = 1, 2, \ldots, P$, is extremely important in practice; with a poor choice of these values, the transformed desired signal vector can be highly distorted. Therefore, the ς_p's should be found in such a way that distortion is minimized. We can rewrite the distortion-based MSE as

$$
\begin{aligned}
J_{\text{ds}}\left(\widetilde{\mathbf{H}}\right) &= \text{tr}\left(\mathbf{\Lambda}_{\widetilde{\mathbf{x}}}\right) + \text{tr}\left(\widetilde{\mathbf{H}}\mathbf{\Phi}_{\mathbf{x}}\widetilde{\mathbf{H}}^H\right) - \text{tr}\left(\widetilde{\mathbf{H}}\mathbf{T}_{\mathbf{x}}\mathbf{\Lambda}_{\widetilde{\mathbf{x}}}\right) - \text{tr}\left(\mathbf{\Lambda}_{\widetilde{\mathbf{x}}}\mathbf{T}_{\mathbf{x}}^H\widetilde{\mathbf{H}}^H\right) \\
&= \text{tr}\left(\mathbf{\Lambda}_{\widetilde{\mathbf{x}}}\right) + \sum_{p=1}^{P} \widetilde{\mathbf{h}}_p^H \mathbf{\Phi}_{\mathbf{x}} \widetilde{\mathbf{h}}_p - \sum_{p=1}^{P} \lambda_{\mathbf{x},p} \widetilde{\mathbf{h}}_p^H \mathbf{q}_{\mathbf{x},p} - \sum_{p=1}^{P} \lambda_{\mathbf{x},p} \mathbf{q}_{\mathbf{x},p}^H \widetilde{\mathbf{h}}_p.
\end{aligned}
\tag{2.48}
$$

Substituting (2.44) into (2.48), we get

$$
\begin{aligned}
J_{\text{ds}}\left(\widetilde{\mathbf{H}}_{\text{max}}\right) = \text{tr}\left(\mathbf{\Lambda}_{\widetilde{\mathbf{x}}}\right) + \mathbf{b}_1^H \mathbf{\Phi}_{\mathbf{x}} \mathbf{b}_1 \sum_{p=1}^{P} \left|\varsigma_p\right|^2 - \sum_{p=1}^{P} \varsigma_p \lambda_{\mathbf{x},p} \mathbf{b}_1^H \mathbf{q}_{\mathbf{x},p} \\
- \sum_{p=1}^{P} \varsigma_p^* \lambda_{\mathbf{x},p} \mathbf{q}_{\mathbf{x},p}^H \mathbf{b}_1,
\end{aligned}
\tag{2.49}
$$

where the superscript $*$ is the complex conjugation. Minimizing (2.49) with respect to the ς_p's, we find

$$
\begin{aligned}
\varsigma_p &= \frac{\lambda_{\mathbf{x},p} \mathbf{q}_{\mathbf{x},p}^H \mathbf{b}_1}{\mathbf{b}_1^H \mathbf{\Phi}_{\mathbf{x}} \mathbf{b}_1} \\
&= \frac{\lambda_{\mathbf{x},p} \mathbf{q}_{\mathbf{x},p}^H \mathbf{b}_1}{\lambda_1}, \quad p = 1, 2, \ldots, P,
\end{aligned}
\tag{2.50}
$$

where $\lambda_1 = \mathbf{b}_1^H \mathbf{\Phi}_{\mathbf{x}} \mathbf{b}_1$ (see Chapter 3). Plugging these optimal values in (2.44), we obtain the optimal maximum SNR filtering matrix with minimum desired signal distortion:

$$
\widetilde{\mathbf{H}}_{\text{max}} = \mathbf{\Lambda}_{\widetilde{\mathbf{x}}} \mathbf{T}_{\mathbf{x}}^H \frac{\mathbf{b}_1 \mathbf{b}_1^H}{\lambda_1}.
\tag{2.51}
$$

We also deduce that the maximum SNR filtering matrix for the estimation of $\mathbf{x}$ is

$$
\begin{aligned}
\mathbf{H}_{\text{max}} &= \mathbf{T}_{\mathbf{x}} \widetilde{\mathbf{H}}_{\text{max}} \\
&= \mathbf{\Phi}_{\mathbf{x}} \frac{\mathbf{b}_1 \mathbf{b}_1^H}{\lambda_1}.
\end{aligned}
\tag{2.52}
$$

2.4.2 Wiener

If we differentiate the MSE criterion, $J\left(\widetilde{\mathbf{H}}\right)$, with respect to $\widetilde{\mathbf{H}}$ and equate the result to zero, we find the Wiener filtering matrix:

$$\begin{aligned}\widetilde{\mathbf{H}}_{\mathrm{W}} &= \mathbf{\Lambda}_{\widetilde{\mathbf{x}}}\mathbf{T}_{\mathbf{x}}^{H}\mathbf{\Phi}_{\mathbf{y}}^{-1} \\ &= \mathbf{T}_{\mathbf{x}}^{H}\mathbf{\Phi}_{\mathbf{x}}\mathbf{\Phi}_{\mathbf{y}}^{-1} \\ &= \mathbf{T}_{\mathbf{x}}^{H}\left(\mathbf{I}_{M} - \mathbf{\Phi}_{\mathbf{v}}\mathbf{\Phi}_{\mathbf{y}}^{-1}\right). \end{aligned} \tag{2.53}$$

We deduce that the equivalent Wiener filtering matrix for the estimation of the vector $\mathbf{x}$ is

$$\begin{aligned}\mathbf{H}_{\mathrm{W}} &= \mathbf{T}_{\mathbf{x}}\widetilde{\mathbf{H}}_{\mathrm{W}} \\ &= \mathbf{T}_{\mathbf{x}}\mathbf{\Lambda}_{\widetilde{\mathbf{x}}}\mathbf{T}_{\mathbf{x}}^{H}\mathbf{\Phi}_{\mathbf{y}}^{-1} \\ &= \mathbf{\Phi}_{\mathbf{x}}\mathbf{\Phi}_{\mathbf{y}}^{-1} \\ &= \mathbf{I}_{M} - \mathbf{\Phi}_{\mathbf{v}}\mathbf{\Phi}_{\mathbf{y}}^{-1}, \end{aligned} \tag{2.54}$$

which corresponds to the classical Wiener filtering matrix [2]. It is extremely important to observe that, thanks to the eigenvalue decomposition and the nullspace of $\mathbf{\Phi}_{\mathbf{x}}$, the size $(P \times M)$ of the proposed Wiener filtering matrix is smaller than the size $(M \times M)$ of the classical Wiener filtering matrix, for the estimation of the desired signal vector $\mathbf{x}$, while the two methods lead to the exactly same result. We deduce that the optimal Wiener filter for the estimation of the mth component of $\mathbf{x}$ is

$$\begin{aligned}\mathbf{h}_{\mathrm{W},m} &= \mathbf{\Phi}_{\mathbf{y}}^{-1}\mathbf{\Phi}_{\mathbf{x}}\mathbf{i}_{m} \\ &= \left(\mathbf{I}_{M} - \mathbf{\Phi}_{\mathbf{y}}^{-1}\mathbf{\Phi}_{\mathbf{v}}\right)\mathbf{i}_{m}, \end{aligned} \tag{2.55}$$

where $\mathbf{i}_m$ is the mth column of $\mathbf{I}_M$.

By applying the Woodbury's identity in (2.11) and then substituting the result in (2.53), we easily deduce another form of the Wiener filtering matrix:

$$\begin{aligned}\widetilde{\mathbf{H}}_{\mathrm{W}} &= \left(\mathbf{I}_{P} + \mathbf{\Lambda}_{\widetilde{\mathbf{x}}}\mathbf{T}_{\mathbf{x}}^{H}\mathbf{\Phi}_{\mathbf{v}}^{-1}\mathbf{T}_{\mathbf{x}}\right)^{-1}\mathbf{\Lambda}_{\widetilde{\mathbf{x}}}\mathbf{T}_{\mathbf{x}}^{H}\mathbf{\Phi}_{\mathbf{v}}^{-1} \\ &= \left(\mathbf{\Lambda}_{\widetilde{\mathbf{x}}}^{-1} + \mathbf{T}_{\mathbf{x}}^{H}\mathbf{\Phi}_{\mathbf{v}}^{-1}\mathbf{T}_{\mathbf{x}}\right)^{-1}\mathbf{T}_{\mathbf{x}}^{H}\mathbf{\Phi}_{\mathbf{v}}^{-1}. \end{aligned} \tag{2.56}$$

The previous expression is interesting because it shows an obvious link with some other optimal rectangular filtering matrices as it will be verified later. We also have

$$\mathbf{H}_{\mathrm{W}} = \mathbf{T}_{\mathbf{x}} \left(\mathbf{I}_P + \mathbf{\Lambda}_{\widetilde{\mathbf{x}}} \mathbf{T}_{\mathbf{x}}^H \mathbf{\Phi}_{\mathbf{v}}^{-1} \mathbf{T}_{\mathbf{x}} \right)^{-1} \mathbf{\Lambda}_{\widetilde{\mathbf{x}}} \mathbf{T}_{\mathbf{x}}^H \mathbf{\Phi}_{\mathbf{v}}^{-1}. \tag{2.57}$$

If $\mathbf{\Phi}_{\mathbf{v}}$ is diagonal, i.e., $\mathbf{\Phi}_{\mathbf{v}} = \sigma_{\mathrm{wn}}^2 \mathbf{I}_M$, (2.57) simplifies to

$$\mathbf{H}_{\mathrm{W}} = \mathbf{T}_{\mathbf{x}} \left(\sigma_{\mathrm{wn}}^2 \mathbf{I}_P + \mathbf{\Lambda}_{\widetilde{\mathbf{x}}} \right)^{-1} \mathbf{\Lambda}_{\widetilde{\mathbf{x}}} \mathbf{T}_{\mathbf{x}}^H. \tag{2.58}$$

This shows how the speech subspace is modified to get a good estimate of $\mathbf{x}$ from $\mathbf{y}$ with Wiener.

Property 2.1. *The output SNR with the Wiener filtering matrix is always greater than or equal to the input SNR, i.e.,* $\mathrm{oSNR}\left(\widetilde{\mathbf{H}}_{\mathrm{W}}\right) \geq \mathrm{iSNR}$.

Obviously, we have

$$\mathrm{oSNR}\left(\widetilde{\mathbf{H}}_{\mathrm{W}}\right) \leq \mathrm{oSNR}\left(\widetilde{\mathbf{H}}_{\max}\right) \tag{2.59}$$

and, in general,

$$\upsilon_{\mathrm{sd}}\left(\widetilde{\mathbf{H}}_{\mathrm{W}}\right) \leq \upsilon_{\mathrm{sd}}\left(\widetilde{\mathbf{H}}_{\max}\right). \tag{2.60}$$

2.4.3 MVDR

The celebrated minimum variance distortionless response (MVDR) filter proposed by Capon [5, 6] has the ability to reduce the noise without distorting the desired signal. It is derived by simply minimizing the MSE of the residual noise, $J_{\mathrm{rs}}\left(\widetilde{\mathbf{H}}\right)$, with the constraint that the desired signal is not distorted. Mathematically, this is equivalent to

$$\min_{\widetilde{\mathbf{H}}} \mathrm{tr}\left(\widetilde{\mathbf{H}} \mathbf{\Phi}_{\mathbf{v}} \widetilde{\mathbf{H}}^H \right) \quad \text{subject to} \quad \widetilde{\mathbf{H}} \mathbf{T}_{\mathbf{x}} = \mathbf{I}_P. \tag{2.61}$$

The solution to the above optimization problem is

$$\widetilde{\mathbf{H}}_{\mathrm{MVDR}} = \left(\mathbf{T}_{\mathbf{x}}^H \mathbf{\Phi}_{\mathbf{v}}^{-1} \mathbf{T}_{\mathbf{x}} \right)^{-1} \mathbf{T}_{\mathbf{x}}^H \mathbf{\Phi}_{\mathbf{v}}^{-1}, \tag{2.62}$$

which is interesting to compare to $\widetilde{\mathbf{H}}_{\mathrm{W}}$ [eq. (2.56)]. We deduce that the MVDR for the estimation of $\mathbf{x}$ is

$$\mathbf{H}_{\mathrm{MVDR}} = \mathbf{T}_{\mathbf{x}} \left(\mathbf{T}_{\mathbf{x}}^H \mathbf{\Phi}_{\mathbf{v}}^{-1} \mathbf{T}_{\mathbf{x}} \right)^{-1} \mathbf{T}_{\mathbf{x}}^H \mathbf{\Phi}_{\mathbf{v}}^{-1}. \tag{2.63}$$

Of course, for $P = M$, the MVDR filtering matrix simplifies to the identity matrix, i.e., $\mathbf{H}_{\mathrm{MVDR}} = \mathbf{I}_M$. As a consequence, we can state that the higher the dimension of the nullspace of $\mathbf{\Phi}_{\mathbf{x}}$, the more the MVDR is efficient in terms of noise reduction. The best scenario corresponds to $P = 1$. If $\mathbf{\Phi}_{\mathbf{v}} = \sigma_{\text{wn}}^2 \mathbf{I}_M$, the MVDR simplifies to

$$\mathbf{H}_{\text{MVDR}} = \mathbf{T}_{\mathbf{x}} \mathbf{T}_{\mathbf{x}}^H. \tag{2.64}$$

In this case, speech enhancement consists of projecting $\mathbf{y}$ onto the speech subspace.

Obviously, with the MVDR filtering matrix, we have no distortion, i.e.,

$$\xi_{\text{sr}} \left(\widetilde{\mathbf{H}}_{\text{MVDR}} \right) = 1, \tag{2.65}$$

$$\upsilon_{\text{sd}} \left(\widetilde{\mathbf{H}}_{\text{MVDR}} \right) = 0. \tag{2.66}$$

Using the Woodbury's identity, we can rewrite the MVDR filtering matrix as

$$\widetilde{\mathbf{H}}_{\text{MVDR}} = \left(\mathbf{T}_{\mathbf{x}}^H \mathbf{\Phi}_{\mathbf{y}}^{-1} \mathbf{T}_{\mathbf{x}} \right)^{-1} \mathbf{T}_{\mathbf{x}}^H \mathbf{\Phi}_{\mathbf{y}}^{-1}. \tag{2.67}$$

From (2.67), we deduce the relationship between the MVDR and Wiener filtering matrices:

$$\widetilde{\mathbf{H}}_{\text{MVDR}} = \left(\widetilde{\mathbf{H}}_{\text{W}} \mathbf{T}_{\mathbf{x}} \right)^{-1} \widetilde{\mathbf{H}}_{\text{W}}. \tag{2.68}$$

Expression (2.67) can also be derived from the following reasoning. We know that

$$\mathbf{x} = \mathbf{T}_{\mathbf{x}} \widetilde{\mathbf{x}}, \tag{2.69}$$

where $\mathbf{T}_{\mathbf{x}}$ can be seen as a temporal prediction matrix. Left multiplying both sides of the previous expression by $\widetilde{\mathbf{H}}$, we see that the distortionless constraint is $\widetilde{\mathbf{H}} \mathbf{T}_{\mathbf{x}} = \mathbf{I}_P$. Now, by minimizing the energy at the output of the filtering matrix, i.e., $\text{tr} \left(\widetilde{\mathbf{H}} \mathbf{\Phi}_{\mathbf{y}} \widetilde{\mathbf{H}}^H \right)$, with the distortionless constraint, we find (2.67).

Property 2.2. *The output SNR with the MVDR filtering matrix is always greater than or equal to the input SNR, i.e.,* $\text{oSNR} \left(\widetilde{\mathbf{H}}_{\text{MVDR}} \right) \geq \text{iSNR}$.

We have

$$\text{oSNR} \left(\widetilde{\mathbf{H}}_{\text{MVDR}} \right) \leq \text{oSNR} \left(\widetilde{\mathbf{H}}_{\text{W}} \right) \leq \text{oSNR} \left(\widetilde{\mathbf{H}}_{\max} \right). \tag{2.70}$$

2.4.4 Tradeoff

In the tradeoff approach [2, 3], we minimize the speech distortion index with the constraint that the noise reduction factor is equal to a positive value that is greater than 1. Mathematically, this is equivalent to

$$\min_{\widetilde{\mathbf{H}}} J_{\text{ds}}\left(\widetilde{\mathbf{H}}\right) \quad \text{subject to} \quad J_{\text{rs}}\left(\widetilde{\mathbf{H}}\right) = \beta \text{tr}\left(\mathbf{\Phi}_{\mathbf{v}}\right), \tag{2.71}$$

where $0 < \beta < 1$ to insure that we get some noise reduction. By using a Lagrange multiplier, $\mu > 0$, to adjoin the constraint to the cost function and assuming that the matrix $\mathbf{T}_{\mathbf{x}}\mathbf{\Phi}_{\widetilde{\mathbf{x}}}\mathbf{T}_{\mathbf{x}}^H + \mu\mathbf{\Phi}_{\mathbf{v}}$ is invertible, we easily deduce the tradeoff filtering matrix:

$$\widetilde{\mathbf{H}}_{\text{T},\mu} = \mathbf{\Lambda}_{\widetilde{\mathbf{x}}}\mathbf{T}_{\mathbf{x}}^H\left(\mathbf{T}_{\mathbf{x}}\mathbf{\Lambda}_{\widetilde{\mathbf{x}}}\mathbf{T}_{\mathbf{x}}^H + \mu\mathbf{\Phi}_{\mathbf{v}}\right)^{-1}, \tag{2.72}$$

which can be rewritten, thanks to the Woodbury's identity, as

$$\widetilde{\mathbf{H}}_{\text{T},\mu} = \left(\mu\mathbf{\Lambda}_{\widetilde{\mathbf{x}}}^{-1} + \mathbf{T}_{\mathbf{x}}^H\mathbf{\Phi}_{\mathbf{v}}^{-1}\mathbf{T}_{\mathbf{x}}\right)^{-1}\mathbf{T}_{\mathbf{x}}^H\mathbf{\Phi}_{\mathbf{v}}^{-1}, \tag{2.73}$$

where μ satisfies $J_{\text{rs}}\left(\widetilde{\mathbf{H}}_{\text{T},\mu}\right) = \beta \text{tr}\left(\mathbf{\Phi}_{\mathbf{v}}\right)$. Usually, μ is chosen in a heuristic way, and for

- $\mu = 1$, $\widetilde{\mathbf{H}}_{\text{T},1} = \widetilde{\mathbf{H}}_{\text{W}}$, which is the Wiener filtering matrix;
- $\mu = 0$ [from (2.73)], $\widetilde{\mathbf{H}}_{\text{T},0} = \widetilde{\mathbf{H}}_{\text{MVDR}}$, which is the MVDR filtering matrix;
- $\mu > 1$, results in a filtering matrix with low residual noise at the expense of high speech distortion (as compared to Wiener);
- $\mu < 1$, results in a filtering matrix with high residual noise and low speech distortion (as compared to Wiener).

Property 2.3. *The output SNR with the tradeoff filtering matrix is always greater than or equal to the input SNR, i.e.,* $\text{oSNR}\left(\widetilde{\mathbf{H}}_{\text{T},\mu}\right) \geq \text{iSNR}, \forall \mu \geq 0$.

We should have, for $\mu \geq 1$,

$$\text{oSNR}\left(\widetilde{\mathbf{H}}_{\text{MVDR}}\right) \leq \text{oSNR}\left(\widetilde{\mathbf{H}}_{\text{W}}\right) \leq \text{oSNR}\left(\widetilde{\mathbf{H}}_{\text{T},\mu}\right) \leq \text{oSNR}\left(\widetilde{\mathbf{H}}_{\text{max}}\right), \tag{2.74}$$

$$0 = \upsilon_{\text{sd}}\left(\widetilde{\mathbf{H}}_{\text{MVDR}}\right) \leq \upsilon_{\text{sd}}\left(\widetilde{\mathbf{H}}_{\text{W}}\right) \leq \upsilon_{\text{sd}}\left(\widetilde{\mathbf{H}}_{\text{T},\mu}\right), \tag{2.75}$$

and, for $\mu \leq 1$,

$$\text{oSNR}\left(\widetilde{\mathbf{H}}_{\text{MVDR}}\right) \leq \text{oSNR}\left(\widetilde{\mathbf{H}}_{\text{T},\mu}\right) \leq \text{oSNR}\left(\widetilde{\mathbf{H}}_{\text{W}}\right) \leq \text{oSNR}\left(\widetilde{\mathbf{H}}_{\text{max}}\right), \tag{2.76}$$

$$0 = \upsilon_{\text{sd}}\left(\widetilde{\mathbf{H}}_{\text{MVDR}}\right) \leq \upsilon_{\text{sd}}\left(\widetilde{\mathbf{H}}_{\text{T},\mu}\right) \leq \upsilon_{\text{sd}}\left(\widetilde{\mathbf{H}}_{\text{W}}\right). \tag{2.77}$$

Let us end this subsection by writing the tradeoff filtering matrix for the estimation of $\mathbf{x}$:

$$\mathbf{H}_{\text{T},\mu} = \mathbf{T}_{\mathbf{x}}\left(\mu\mathbf{\Lambda}_{\widetilde{\mathbf{x}}}^{-1} + \mathbf{T}_{\mathbf{x}}^{H}\mathbf{\Phi}_{\mathbf{v}}^{-1}\mathbf{T}_{\mathbf{x}}\right)^{-1}\mathbf{T}_{\mathbf{x}}^{H}\mathbf{\Phi}_{\mathbf{v}}^{-1}, \tag{2.78}$$

which clearly shows how the speech subspace should be modified in order to make a compromise between noise reduction and speech distortion.

2.4.5 LCMV

Let us first decompose the noise correlation matrix as

$$\mathbf{\Phi}_{\mathbf{v}} = \mathbf{Q}_{\mathbf{v}}\mathbf{\Lambda}_{\mathbf{v}}\mathbf{Q}_{\mathbf{v}}^{H}, \tag{2.79}$$

where the unitary and diagonal matrices $\mathbf{Q}_{\mathbf{v}}$ and $\mathbf{\Lambda}_{\mathbf{v}}$ are defined similarly to $\mathbf{Q}_{\mathbf{x}}$ and $\mathbf{\Lambda}_{\mathbf{x}}$, respectively. We assume that the (positive) eigenvalues of $\mathbf{\Phi}_{\mathbf{v}}$ have the following structure: $\lambda_{\mathbf{v},1} \geq \lambda_{\mathbf{v},2} \geq \cdots \geq \lambda_{\mathbf{v},Q} > \sigma_{\text{wn}}^2$ and $\lambda_{\mathbf{v},Q+1} = \lambda_{\mathbf{v},Q+2} = \cdots = \lambda_{\mathbf{v},M} = \sigma_{\text{wn}}^2$, where $P + Q \leq M$. In this case, we can rewrite the unitary matrix as

$$\mathbf{Q}_{\mathbf{v}} = \begin{bmatrix} \mathbf{T}_{\mathbf{v}} & \mathbf{\Xi}_{\mathbf{v}} \end{bmatrix}, \tag{2.80}$$

where the $M \times Q$ matrix $\mathbf{T}_{\mathbf{v}}$ contains the eigenvectors corresponding to the first Q eigenvalues of $\mathbf{\Phi}_{\mathbf{v}}$ and the $M \times (M - Q)$ matrix $\mathbf{\Xi}_{\mathbf{v}}$ contains the eigenvectors corresponding to the last $M - Q$ eigenvalues of $\mathbf{\Phi}_{\mathbf{v}}$. As a result, the noise signal vector can be decomposed as

$$\mathbf{v} = \mathbf{v}_{\text{c}} + \mathbf{v}_{\text{u}}, \tag{2.81}$$

where

$$\mathbf{v}_{\text{c}} = \mathbf{T}_{\mathbf{v}}\mathbf{T}_{\mathbf{v}}^{H}\mathbf{v} \tag{2.82}$$

corresponds to the correlated noise,

$$\mathbf{v}_{\text{u}} = \mathbf{\Xi}_{\mathbf{v}}\mathbf{\Xi}_{\mathbf{v}}^{H}\mathbf{v} \tag{2.83}$$

corresponds to the uncorrelated noise, and $E\left(\mathbf{v}_{\text{c}}\mathbf{v}_{\text{u}}^{H}\right) = \mathbf{0}_{M \times M}$.

The LCMV approach [7,8] consists of estimating $\mathbf{x}$ without any distortion, completely removing the correlated noise, and attenuating the uncorrelated noise as much as possible. It follows that the constraints are

$$\widetilde{\mathbf{H}}\mathbf{C}_{\mathbf{xv}} = \left[\mathbf{I}_P \ \mathbf{0}_{P\times Q}\right], \tag{2.84}$$

where

$$\mathbf{C}_{\mathbf{xv}} = \left[\mathbf{T}_{\mathbf{x}} \ \mathbf{T}_{\mathbf{v}}\right] \tag{2.85}$$

is the constraint matrix of size $M \times (P + Q)$. The optimization problem is now

$$\min_{\widetilde{\mathbf{H}}} \mathrm{tr}\left(\widetilde{\mathbf{H}}\mathbf{\Phi}_{\mathbf{y}}\widetilde{\mathbf{H}}^H\right) \quad \text{subject to} \quad \widetilde{\mathbf{H}}\mathbf{C}_{\mathbf{xv}} = \left[\mathbf{I}_P \ \mathbf{0}_{P\times Q}\right], \tag{2.86}$$

from which we find the LCMV filtering matrix:

$$\widetilde{\mathbf{H}}_{\mathrm{LCMV}} = \left[\mathbf{I}_P \ \mathbf{0}_{P\times Q}\right]\left(\mathbf{C}_{\mathbf{xv}}^H\mathbf{\Phi}_{\mathbf{y}}^{-1}\mathbf{C}_{\mathbf{xv}}\right)^{-1}\mathbf{C}_{\mathbf{xv}}^H\mathbf{\Phi}_{\mathbf{y}}^{-1}. \tag{2.87}$$

We immediately see from (2.87) that we must have $P + Q \leq M$, otherwise the matrix $\mathbf{C}_{\mathbf{xv}}^H\mathbf{\Phi}_{\mathbf{y}}^{-1}\mathbf{C}_{\mathbf{xv}}$ is not invertible. For $P + Q > M$, the LCMV filter does not exist. For $P + Q = M$, the LCMV simplifies to

$$\widetilde{\mathbf{H}}_{\mathrm{LCMV}} = \left[\mathbf{I}_P \ \mathbf{0}_{P\times Q}\right]\mathbf{C}_{\mathbf{xv}}^{-1}. \tag{2.88}$$

Finally, we see that the LCMV filter for the estimation of $\mathbf{x}$ is

$$\mathbf{H}_{\mathrm{LCMV}} = \mathbf{T}_{\mathbf{x}}\left[\mathbf{I}_P \ \mathbf{0}_{P\times Q}\right]\left(\mathbf{C}_{\mathbf{xv}}^H\mathbf{\Phi}_{\mathbf{y}}^{-1}\mathbf{C}_{\mathbf{xv}}\right)^{-1}\mathbf{C}_{\mathbf{xv}}^H\mathbf{\Phi}_{\mathbf{y}}^{-1}. \tag{2.89}$$

REFERENCES

[1] G.H. Golub, C.F. Van Loan, Matrix Computations, third ed., The Johns Hopkins University Press, Baltimore, Maryland, 1996.

[2] J. Benesty, J. Chen, Y. Huang, I. Cohen, Noise Reduction in Speech Processing, Springer-Verlag, Berlin, Germany, 2009.

[3] J. Benesty, J. Chen, Optimal Time-domain Noise Reduction Filters—A Theoretical Study, Springer Briefs in Electrical and Computer Engineering, 2011.

[4] J. Benesty, S. Makino, J. Chen (Eds.), Speech Enhancement, Springer-Verlag, Berlin, Germany, 2005.

[5] J. Capon, High resolution frequency-wavenumber spectrum analysis, Proc. IEEE 57 (1969) 1408–1418.

[6] R.T. Lacoss, Data adaptive spectral analysis methods, Geophysics 36 (1971) 661–675.

[7] O. Frost, An algorithm for linearly constrained adaptive array processing, Proc. IEEE 60 (1972) 926–935.

[8] M. Er, A. Cantoni, Derivative constraints for broad-band element space antenna array processors, IEEE Trans. Acoust. Speech Signal Process. 31 (1983) 1378–1393.

CHAPTER 3

General Concept with the Joint Diagonalization of the Speech and Noise Correlation Matrices

In the previous chapter, we showed how the eigenvalue decomposition of the speech correlation matrix can be exploited in the derivation of different types of optimal filtering matrices for the general problem of speech enhancement. This chapter attempts to show the same results but with the joint diagonalization of the speech and noise correlation matrices. We will see that there are some subtle differences between these two approaches with many more possibilities with joint diagonalization, suggesting that this tool is very natural to use in this problem.

3.1 SIGNAL MODEL AND PROBLEM FORMULATION

We consider the same signal model as the one presented in Chapter 2, i.e., $\mathbf{y} = \mathbf{x} + \mathbf{v}$, where the correlation matrix of $\mathbf{y}$ is $\mathbf{\Phi}_\mathbf{y} = \mathbf{\Phi}_\mathbf{x} + \mathbf{\Phi}_\mathbf{v}$. Again, it is assumed that the rank of the speech correlation matrix, $\mathbf{\Phi}_\mathbf{x}$, is equal to $P \leq M$ and the rank of the noise correlation matrix, $\mathbf{\Phi}_\mathbf{v}$, is equal to M.

The use of the joint diagonalization was first proposed in [1] and then in [2]. In this part, we show how it can be applied to our signal model in a more general and rigorous way in order to extract the desired signal, $\mathbf{x}$.

The two Hermitian matrices $\mathbf{\Phi}_\mathbf{x}$ and $\mathbf{\Phi}_\mathbf{v}$ can be jointly diagonalized as follows [3]:

$$\mathbf{B}^H \mathbf{\Phi}_\mathbf{x} \mathbf{B} = \mathbf{\Lambda}, \tag{3.1}$$

$$\mathbf{B}^H \mathbf{\Phi}_\mathbf{v} \mathbf{B} = \mathbf{I}_M, \tag{3.2}$$

where $\mathbf{B}$ is a full-rank square matrix (of size $M \times M$) and $\mathbf{\Lambda}$ is a diagonal matrix whose main elements are real and nonnegative. Furthermore, $\mathbf{\Lambda}$ and $\mathbf{B}$ are the eigenvalue and eigenvector matrices, respectively, of $\mathbf{\Phi}_\mathbf{v}^{-1}\mathbf{\Phi}_\mathbf{x}$, i.e.,

$$\mathbf{\Phi}_\mathbf{v}^{-1}\mathbf{\Phi}_\mathbf{x}\mathbf{B} = \mathbf{B}\mathbf{\Lambda}. \tag{3.3}$$

Since the rank of the matrix $\mathbf{\Phi}_\mathbf{x}$ is equal to P, the eigenvalues of $\mathbf{\Phi}_\mathbf{v}^{-1}\mathbf{\Phi}_\mathbf{x}$ can be ordered as $\lambda_1 \geq \lambda_2 \geq \cdots \geq \lambda_P > \lambda_{P+1} = \cdots = \lambda_M = 0$. In

Speech Enhancement. http://dx.doi.org/10.1016/B978-0-12-800139-4.00003-7

other words, the last $M - P$ eigenvalues of the matrix product $\mathbf{\Phi}_\mathbf{v}^{-1}\mathbf{\Phi}_\mathbf{x}$ are exactly zero, while its first P eigenvalues are positive, with λ_1 being the maximum eigenvalue. We also denote by $\mathbf{b}_1, \mathbf{b}_2, \ldots, \mathbf{b}_P, \mathbf{b}_{P+1}, \ldots, \mathbf{b}_M$, the corresponding eigenvectors. A consequence of this joint diagonalization is that the noisy signal correlation matrix can also be diagonalized as

$$\mathbf{B}^H \mathbf{\Phi}_\mathbf{y} \mathbf{B} = \mathbf{\Lambda} + \mathbf{I}_M. \tag{3.4}$$

We can decompose the matrix $\mathbf{B}$ as

$$\mathbf{B} = \left[\mathbf{T} \ \mathbf{\Xi} \right], \tag{3.5}$$

where

$$\mathbf{T} = \left[\mathbf{b}_1 \ \mathbf{b}_2 \cdots \mathbf{b}_P \right] \tag{3.6}$$

is an $M \times P$ matrix that spans the speech-plus-noise subspace and

$$\mathbf{\Xi} = \left[\mathbf{b}_{P+1} \ \mathbf{b}_{P+2} \cdots \mathbf{b}_M \right] \tag{3.7}$$

is an $M \times (M - P)$ matrix that spans the noise subspace.

Let us define the matrix:

$$\begin{aligned} \mathbf{B}' &= \mathbf{\Phi}_\mathbf{v}^{1/2} \mathbf{B} \\ &= \left[\mathbf{b}_1' \ \mathbf{b}_2' \cdots \mathbf{b}_M' \right]. \end{aligned} \tag{3.8}$$

Expression (3.2) becomes

$$\mathbf{B}'^H \mathbf{B}' = \mathbf{I}_M. \tag{3.9}$$

Since $\mathbf{B}'$ is a full-rank matrix, we have

$$\mathbf{B}'^H = \mathbf{B}'^{-1} \tag{3.10}$$

and

$$\mathbf{B}' \mathbf{B}'^H = \mathbf{I}_M. \tag{3.11}$$

As a result, we deduce that

$$\begin{aligned} \mathbf{I}_M &= \mathbf{\Phi}_\mathbf{v}^{1/2} \mathbf{T} \mathbf{T}^H \mathbf{\Phi}_\mathbf{v}^{1/2} + \mathbf{\Phi}_\mathbf{v}^{1/2} \mathbf{\Xi} \mathbf{\Xi}^H \mathbf{\Phi}_\mathbf{v}^{1/2} \\ &= \mathbf{T}' \mathbf{T}'^H + \mathbf{\Xi}' \mathbf{\Xi}'^H, \end{aligned} \tag{3.12}$$

where

$$\mathbf{T}' = \mathbf{\Phi}_\mathbf{v}^{1/2} \mathbf{T}, \tag{3.13}$$

$$\Xi' = \mathbf{\Phi}_{\mathbf{v}}^{1/2}\Xi. \tag{3.14}$$

It is easy to verify that $\mathbf{T}'\mathbf{T}'^{H}$ and $\Xi'\Xi'^{H}$ are two orthogonal projection matrices of rank P and $M - P$, respectively.

Now, we define our new signal model as

$$\begin{aligned}\mathbf{y}' &= \mathbf{\Phi}_{\mathbf{v}}^{-1/2}\mathbf{y} \\ &= \mathbf{x}' + \mathbf{v}',\end{aligned} \tag{3.15}$$

where $\mathbf{x}' = \mathbf{\Phi}_{\mathbf{v}}^{-1/2}\mathbf{x}$ and $\mathbf{v}' = \mathbf{\Phi}_{\mathbf{v}}^{-1/2}\mathbf{v}$ are the new speech and noise signal vectors, respectively. When $\mathbf{x}'$ is estimated correctly, it is then easy to deduce $\mathbf{x}$ from this estimator. Furthermore, $\mathbf{T}'\mathbf{T}'^{H}$ is the orthogonal projector onto the new speech-plus-noise subspace while $\Xi'\Xi'^{H}$ is the orthogonal projector onto the new noise subspace. Using (3.12), we can write the new speech vector as

$$\begin{aligned}\mathbf{x}' &= \mathbf{B}'\mathbf{B}'^{H}\mathbf{x}' \\ &= \mathbf{T}'\mathbf{T}'^{H}\mathbf{x}' \\ &= \mathbf{T}'\widetilde{\mathbf{x}}',\end{aligned} \tag{3.16}$$

where

$$\widetilde{\mathbf{x}}' = \mathbf{T}'^{H}\mathbf{x}' \tag{3.17}$$

is the transformed desired signal vector of length P. Therefore, with joint diagonalization, the signal model for noise reduction is

$$\mathbf{y}' = \mathbf{T}'\widetilde{\mathbf{x}}' + \mathbf{v}'. \tag{3.18}$$

Thanks to this transformation, there is no need to use the noise subspace Ξ'. The correlation matrix of $\mathbf{y}'$ is then

$$\mathbf{\Phi}_{\mathbf{y}'} = \mathbf{T}'\mathbf{\Phi}_{\widetilde{\mathbf{x}}'}\mathbf{T}'^{H} + \mathbf{I}_M, \tag{3.19}$$

where

$$\begin{aligned}\mathbf{\Phi}_{\widetilde{\mathbf{x}}'} &= E\left(\widetilde{\mathbf{x}}'\widetilde{\mathbf{x}}'^{H}\right) \\ &= \mathbf{T}'^{H}\mathbf{\Phi}_{\mathbf{x}'}\mathbf{T}' \\ &= \mathbf{T}'^{H}\mathbf{\Phi}_{\mathbf{v}}^{-1/2}\mathbf{\Phi}_{\mathbf{x}}\mathbf{\Phi}_{\mathbf{v}}^{-1/2}\mathbf{T}' \\ &= \mathrm{diag}\left(\lambda_1, \lambda_2, \ldots, \lambda_P\right) \\ &= \mathbf{\Lambda}_{\widetilde{\mathbf{x}}'}\end{aligned} \tag{3.20}$$

and $\mathbf{\Phi}_{\mathbf{x}'} = \mathbf{T}'\mathbf{\Lambda}_{\widetilde{\mathbf{x}}'}\mathbf{T}'^{H}$.

3.2 LINEAR FILTERING WITH A RECTANGULAR MATRIX

By applying a linear transformation, $\widetilde{\mathbf{H}}'$ (of size $P \times M$), to the signal vector, $\mathbf{y}'$, we get an estimate of $\widetilde{\mathbf{x}}'$, which can be written as

$$\begin{aligned}\widetilde{\mathbf{z}}' &= \widetilde{\mathbf{H}}'\mathbf{y}' \\ &= \widetilde{\mathbf{H}}'\left(\mathbf{T}'\widetilde{\mathbf{x}}' + \mathbf{v}'\right) \\ &= \widetilde{\mathbf{x}}'_{\text{fd}} + \widetilde{\mathbf{v}}'_{\text{rn}},\end{aligned} \tag{3.21}$$

where

$$\widetilde{\mathbf{x}}'_{\text{fd}} = \widetilde{\mathbf{H}}'\mathbf{T}'\widetilde{\mathbf{x}}' \tag{3.22}$$

is the filtered (transformed) desired signal and

$$\widetilde{\mathbf{v}}'_{\text{rn}} = \widetilde{\mathbf{H}}'\mathbf{v}' \tag{3.23}$$

is the residual noise. The estimate of $\mathbf{x}'$ follows by applying $\mathbf{T}'$ to $\widetilde{\mathbf{z}}'$, i.e.,

$$\begin{aligned}\mathbf{z}' &= \mathbf{T}'\widetilde{\mathbf{z}}' \\ &= \mathbf{T}'\widetilde{\mathbf{H}}'\mathbf{y}' \\ &= \mathbf{H}'\mathbf{y}',\end{aligned} \tag{3.24}$$

where

$$\mathbf{H}' = \mathbf{T}'\widetilde{\mathbf{H}}' \tag{3.25}$$

is a filtering matrix of size $M \times M$. Continuing this line of thoughts, we deduce that the estimate of $\mathbf{x}$ is

$$\begin{aligned}\mathbf{z} &= \mathbf{\Phi}_{\mathbf{v}}^{1/2}\mathbf{z}' \\ &= \mathbf{\Phi}_{\mathbf{v}}^{1/2}\mathbf{H}'\mathbf{\Phi}_{\mathbf{v}}^{-1/2}\mathbf{y} \\ &= \mathbf{H}\mathbf{y},\end{aligned} \tag{3.26}$$

where

$$\mathbf{H} = \mathbf{\Phi}_{\mathbf{v}}^{1/2}\mathbf{H}'\mathbf{\Phi}_{\mathbf{v}}^{-1/2} \tag{3.27}$$

is the equivalent filtering matrix of size $M \times M$, which is applied to $\mathbf{y}$ to get the estimate of $\mathbf{x}$. The correlation matrix of $\widetilde{\mathbf{z}}'$ is then

$$\begin{aligned}\mathbf{\Phi}_{\widetilde{\mathbf{z}}'} &= E\left(\widetilde{\mathbf{z}}'\widetilde{\mathbf{z}}'^H\right) \\ &= \mathbf{\Phi}_{\widetilde{\mathbf{x}}'_{\text{fd}}} + \mathbf{\Phi}_{\widetilde{\mathbf{v}}'_{\text{rn}}},\end{aligned} \tag{3.28}$$

where

$$\mathbf{\Phi}_{\widetilde{\mathbf{x}}'_{\text{fd}}} = \widetilde{\mathbf{H}}'\mathbf{T}'\mathbf{\Phi}_{\widetilde{\mathbf{x}}'}\mathbf{T}'^H\widetilde{\mathbf{H}}'^H, \tag{3.29}$$

$$\mathbf{\Phi}_{\widetilde{\mathbf{v}}'_{\mathrm{rn}}} = \widetilde{\mathbf{H}}'\widetilde{\mathbf{H}}'^H. \tag{3.30}$$

We also observe that $\mathbf{\Phi}_{\mathbf{z}'} = \mathbf{T}'\mathbf{\Phi}_{\widetilde{\mathbf{z}}'}\mathbf{T}'^H$ and $\mathrm{tr}\left(\mathbf{\Phi}_{\mathbf{z}'}\right) = \mathrm{tr}\left(\mathbf{\Phi}_{\widetilde{\mathbf{z}}'}\right)$. The correlation matrix of $\widetilde{\mathbf{z}}'$ is helpful in the next section.

3.3 PERFORMANCE MEASURES

In this section, we briefly define all performance measures of interest (with respect to $\mathbf{y}'$) when joint diagonalization is applied.

3.3.1 Noise Reduction

The definition of the input SNR is obtained from (3.19):

$$\begin{aligned} \mathrm{iSNR}' &= \frac{\mathrm{tr}\left(\mathbf{T}'\mathbf{\Lambda}_{\widetilde{\mathbf{x}}'}\mathbf{T}'^H\right)}{\mathrm{tr}\left(\mathbf{I}_M\right)} \\ &= \frac{\sum_{p=1}^{P}\lambda_p}{M}. \end{aligned} \tag{3.31}$$

The output SNR, obtained from (3.28), is defined as

$$\begin{aligned} \mathrm{oSNR}\left(\widetilde{\mathbf{H}}'\right) &= \frac{\mathrm{tr}\left(\mathbf{\Phi}_{\widetilde{\mathbf{x}}'_{\mathrm{fd}}}\right)}{\mathrm{tr}\left(\mathbf{\Phi}_{\widetilde{\mathbf{v}}'_{\mathrm{rn}}}\right)} \\ &= \frac{\mathrm{tr}\left(\widetilde{\mathbf{H}}'\mathbf{T}'\mathbf{\Phi}_{\widetilde{\mathbf{x}}'}\mathbf{T}'^H\widetilde{\mathbf{H}}'^H\right)}{\mathrm{tr}\left(\widetilde{\mathbf{H}}'\widetilde{\mathbf{H}}'^H\right)} \\ &= \mathrm{oSNR}\left(\mathbf{H}'\right). \end{aligned} \tag{3.32}$$

The filtering matrix, $\widetilde{\mathbf{H}}'$, should be designed in such a way that $\mathrm{oSNR}\left(\widetilde{\mathbf{H}}'\right) \geq \mathrm{iSNR}'$.

The noise reduction factor is

$$\begin{aligned} \xi_{\mathrm{nr}}\left(\widetilde{\mathbf{H}}'\right) &= \frac{M}{\mathrm{tr}\left(\widetilde{\mathbf{H}}'\widetilde{\mathbf{H}}'^H\right)} \\ &= \xi_{\mathrm{nr}}\left(\mathbf{H}'\right). \end{aligned} \tag{3.33}$$

The design of $\widetilde{\mathbf{H}}'$ should satisfy that $\xi_{\mathrm{nr}}\left(\widetilde{\mathbf{H}}'\right) \geq 1$.

3.3.2 Speech Distortion

To quantify the distortion added to the transformed desired signal, we define the speech reduction factor as

$$\begin{aligned}\xi_{\text{sr}}\left(\widetilde{\mathbf{H}}'\right) &= \frac{\text{tr}\left(\mathbf{\Phi}_{\mathbf{x}'}\right)}{\text{tr}\left(\mathbf{\Phi}_{\widetilde{\mathbf{x}}'_{\text{fd}}}\right)} \\ &= \frac{\text{tr}\left(\mathbf{\Lambda}_{\widetilde{\mathbf{x}}'}\right)}{\text{tr}\left(\widetilde{\mathbf{H}}'\mathbf{T}'\mathbf{\Lambda}_{\widetilde{\mathbf{x}}'}\mathbf{T}'^H\widetilde{\mathbf{H}}'^H\right)} \\ &= \xi_{\text{sr}}\left(\mathbf{H}'\right).\end{aligned} \tag{3.34}$$

A rectangular filtering matrix that does not affect the transformed desired signal requires the constraint:

$$\widetilde{\mathbf{H}}'\mathbf{T}' = \mathbf{I}_P. \tag{3.35}$$

Hence, $\xi_{\text{sr}}\left(\widetilde{\mathbf{H}}'\right) = 1$ in the absence of distortion and $\xi_{\text{sr}}\left(\widetilde{\mathbf{H}}'\right) > 1$ in the presence of distortion.

It is obvious that we always have

$$\frac{\text{oSNR}\left(\widetilde{\mathbf{H}}'\right)}{\text{iSNR}'} = \frac{\xi_{\text{nr}}\left(\widetilde{\mathbf{H}}'\right)}{\xi_{\text{sr}}\left(\widetilde{\mathbf{H}}'\right)}. \tag{3.36}$$

The distortion can also be measured with the speech distortion index:

$$\begin{aligned}\upsilon_{\text{sd}}\left(\widetilde{\mathbf{H}}'\right) &= \frac{E\left[\left(\widetilde{\mathbf{x}}'_{\text{fd}} - \widetilde{\mathbf{x}}'\right)^H\left(\widetilde{\mathbf{x}}'_{\text{fd}} - \widetilde{\mathbf{x}}'\right)\right]}{\text{tr}\left(\mathbf{\Phi}_{\widetilde{\mathbf{x}}'}\right)} \\ &= \frac{\text{tr}\left[\left(\widetilde{\mathbf{H}}'\mathbf{T}' - \mathbf{I}_P\right)\mathbf{\Lambda}_{\widetilde{\mathbf{x}}'}\left(\widetilde{\mathbf{H}}'\mathbf{T}' - \mathbf{I}_P\right)^H\right]}{\text{tr}\left(\mathbf{\Lambda}_{\widetilde{\mathbf{x}}'}\right)} \\ &= \upsilon_{\text{sd}}\left(\mathbf{H}'\right).\end{aligned} \tag{3.37}$$

For optimal rectangular filtering matrices, we should have

$$0 \le \upsilon_{\text{sd}}\left(\widetilde{\mathbf{H}}'\right) \le 1 \tag{3.38}$$

and a value of $\upsilon_{\text{sd}}\left(\widetilde{\mathbf{H}}'\right)$ close to 0 is preferred.

3.3.3 MSE Criterion

It is clear that the error signal between the estimated and desired signals is

$$\begin{aligned}\widetilde{\mathbf{e}}' &= \widetilde{\mathbf{z}}' - \widetilde{\mathbf{x}}' \\ &= \widetilde{\mathbf{H}}'\mathbf{y}' - \widetilde{\mathbf{x}}' \\ &= \widetilde{\mathbf{e}}'_{\text{ds}} + \widetilde{\mathbf{e}}'_{\text{rs}},\end{aligned} \tag{3.39}$$

where

$$\begin{aligned}\widetilde{\mathbf{e}}'_{\text{ds}} &= \widetilde{\mathbf{x}}'_{\text{fd}} - \widetilde{\mathbf{x}}' \\ &= \left(\widetilde{\mathbf{H}}'\mathbf{T}' - \mathbf{I}_P\right)\widetilde{\mathbf{x}}'\end{aligned} \tag{3.40}$$

represents the signal distortion and

$$\begin{aligned}\widetilde{\mathbf{e}}'_{\text{rs}} &= \widetilde{\mathbf{v}}'_{\text{rn}} \\ &= \widetilde{\mathbf{H}}'\mathbf{v}'\end{aligned} \tag{3.41}$$

represents the residual noise. We deduce that the MSE criterion is

$$\begin{aligned}J\left(\widetilde{\mathbf{H}}'\right) &= \text{tr}\left[E\left(\widetilde{\mathbf{e}}'\widetilde{\mathbf{e}}'^H\right)\right] \qquad (3.42)\\ &= \text{tr}\left(\mathbf{\Lambda}_{\widetilde{\mathbf{x}}'}\right) + \text{tr}\left(\widetilde{\mathbf{H}}'\mathbf{\Phi}_{\mathbf{y}'}\widetilde{\mathbf{H}}'^H\right) - \text{tr}\left(\widetilde{\mathbf{H}}'\mathbf{T}'\mathbf{\Lambda}_{\widetilde{\mathbf{x}}'}\right) - \text{tr}\left(\mathbf{\Lambda}_{\widetilde{\mathbf{x}}'}\mathbf{T}'^H\widetilde{\mathbf{H}}'^H\right) \\ &= \text{tr}\left(\mathbf{\Phi}_{\mathbf{x}'}\right) + \text{tr}\left(\widetilde{\mathbf{H}}'\mathbf{\Phi}_{\mathbf{y}'}\widetilde{\mathbf{H}}'^H\right) - \text{tr}\left(\widetilde{\mathbf{H}}'\mathbf{\Phi}_{\mathbf{x}'}\mathbf{T}'\right) - \text{tr}\left(\mathbf{T}'^H\mathbf{\Phi}_{\mathbf{x}'}\widetilde{\mathbf{H}}'^H\right).\end{aligned}$$

Since $E\left(\widetilde{\mathbf{e}}'_{\text{ds}}\widetilde{\mathbf{e}}'^H_{\text{rs}}\right) = \mathbf{0}_{P\times P}$, $J\left(\widetilde{\mathbf{H}}'\right)$ can also be expressed as

$$\begin{aligned}J\left(\widetilde{\mathbf{H}}'\right) &= \text{tr}\left[E\left(\widetilde{\mathbf{e}}'_{\text{ds}}\widetilde{\mathbf{e}}'^H_{\text{ds}}\right)\right] + \text{tr}\left[E\left(\widetilde{\mathbf{e}}'_{\text{rs}}\widetilde{\mathbf{e}}'^H_{\text{rs}}\right)\right] \\ &= J_{\text{ds}}\left(\widetilde{\mathbf{H}}'\right) + J_{\text{rs}}\left(\widetilde{\mathbf{H}}'\right),\end{aligned} \tag{3.43}$$

where

$$\begin{aligned}J_{\text{ds}}\left(\widetilde{\mathbf{H}}'\right) &= \text{tr}\left[\left(\widetilde{\mathbf{H}}'\mathbf{T}' - \mathbf{I}_P\right)\mathbf{\Lambda}_{\widetilde{\mathbf{x}}'}\left(\widetilde{\mathbf{H}}'\mathbf{T}' - \mathbf{I}_P\right)^H\right] \\ &= \text{tr}\left(\mathbf{\Lambda}_{\widetilde{\mathbf{x}}'}\right)\upsilon_{\text{sd}}\left(\widetilde{\mathbf{H}}'\right)\end{aligned} \tag{3.44}$$

and

$$\begin{aligned}J_{\text{rs}}\left(\widetilde{\mathbf{H}}'\right) &= \text{tr}\left(\widetilde{\mathbf{H}}'\widetilde{\mathbf{H}}'^H\right) \\ &= \frac{\text{tr}\left(\mathbf{\Phi}_{\mathbf{v}'}\right)}{\xi_{\text{nr}}\left(\widetilde{\mathbf{H}}'\right)}.\end{aligned} \tag{3.45}$$

Finally, we have

$$\begin{aligned}\frac{J_{\mathrm{ds}}\left(\widetilde{\mathbf{H}}'\right)}{J_{\mathrm{rs}}\left(\widetilde{\mathbf{H}}'\right)} &= \mathrm{iSNR}' \times \xi_{\mathrm{nr}}\left(\widetilde{\mathbf{H}}'\right) \times \upsilon_{\mathrm{sd}}\left(\widetilde{\mathbf{H}}'\right) \\ &= \mathrm{oSNR}\left(\widetilde{\mathbf{H}}'\right) \times \xi_{\mathrm{sr}}\left(\widetilde{\mathbf{H}}'\right) \times \upsilon_{\mathrm{sd}}\left(\widetilde{\mathbf{H}}'\right).\end{aligned} \tag{3.46}$$

3.4 OPTIMAL RECTANGULAR FILTERING MATRICES

We are now ready to derive some important conventional filtering matrices, which are all equivalent to the ones derived in Chapter 2. Some new filtering matrices are proposed as well that link the classical ones among them in a nice and natural way.

3.4.1 Maximum SNR

It can be shown that the output SNR is upper bounded by λ_1. Therefore,

$$\widetilde{\mathbf{H}}'_{\mathrm{max}} = \begin{bmatrix} \varsigma_1 \mathbf{b}_1'^H \\ \varsigma_2 \mathbf{b}_1'^H \\ \vdots \\ \varsigma_P \mathbf{b}_1'^H \end{bmatrix}, \tag{3.47}$$

where ς_p, $p = 1, 2, \ldots, P$ are arbitrary complex numbers with at least one of them different from 0, corresponds to the maximum SNR filtering matrix since

$$\mathrm{oSNR}\left(\widetilde{\mathbf{H}}'_{\mathrm{max}}\right) = \lambda_1. \tag{3.48}$$

We always have

$$\mathrm{oSNR}\left(\widetilde{\mathbf{H}}'_{\mathrm{max}}\right) \geq \mathrm{iSNR}' \tag{3.49}$$

and

$$0 \leq \mathrm{oSNR}\left(\widetilde{\mathbf{H}}'\right) \leq \mathrm{oSNR}\left(\widetilde{\mathbf{H}}'_{\mathrm{max}}\right), \forall \widetilde{\mathbf{H}}'. \tag{3.50}$$

Choosing properly the values of ς_p, $p = 1, 2, \ldots, P$, is important in practice if we want to avoid a large distortion of the transformed desired signal vector. The best way to find these values is by minimizing distortion. Let us write the distortion-based MSE:

$$\begin{aligned} J_{\rm ds}\left(\widetilde{\mathbf{H}}'\right) &= \text{tr}\left(\mathbf{\Phi}_{\mathbf{x}'}\right) + \text{tr}\left(\widetilde{\mathbf{H}}'\mathbf{\Phi}_{\mathbf{x}'}\widetilde{\mathbf{H}}'^H\right) - \text{tr}\left(\widetilde{\mathbf{H}}'\mathbf{\Phi}_{\mathbf{x}'}\mathbf{T}'\right) - \text{tr}\left(\mathbf{T}'^H\mathbf{\Phi}_{\mathbf{x}'}\widetilde{\mathbf{H}}'^H\right) \\ &= \text{tr}\left(\mathbf{\Phi}_{\mathbf{x}'}\right) + \sum_{p=1}^{P} \widetilde{\mathbf{h}}_p'^H \mathbf{\Phi}_{\mathbf{x}'}\widetilde{\mathbf{h}}_p' - \sum_{p=1}^{P} \widetilde{\mathbf{h}}_p'^H \mathbf{\Phi}_{\mathbf{x}'}\mathbf{b}_p' - \sum_{p=1}^{P} \mathbf{b}_p'^H \mathbf{\Phi}_{\mathbf{x}'}\widetilde{\mathbf{h}}_p', \end{aligned} \tag{3.51}$$

where $\widetilde{\mathbf{h}}_p'^H$ is the pth line of $\widetilde{\mathbf{H}}'$. Now, substituting (3.47) into (3.51), we obtain

$$\begin{aligned} J_{\rm ds}\left(\widetilde{\mathbf{H}}'_{\rm max}\right) = \text{tr}\left(\mathbf{\Phi}_{\mathbf{x}'}\right) &+ \mathbf{b}_1'^H \mathbf{\Phi}_{\mathbf{x}'}\mathbf{b}_1' \sum_{p=1}^{P} \left|\varsigma_p\right|^2 - \sum_{p=1}^{P} \varsigma_p \mathbf{b}_1'^H \mathbf{\Phi}_{\mathbf{x}'}\mathbf{b}_p' \\ &- \sum_{p=1}^{P} \varsigma_p^* \mathbf{b}_p'^H \mathbf{\Phi}_{\mathbf{x}'}\mathbf{b}_1', \end{aligned} \tag{3.52}$$

and minimizing this expression with respect to the ς_p's, we find

$$\varsigma_1 = 1, \tag{3.53}$$

$$\varsigma_p = 0, \quad p \neq 1. \tag{3.54}$$

We deduce that the optimal maximum SNR filtering matrix with minimum transformed desired signal distortion is

$$\widetilde{\mathbf{H}}'_{\rm max} = \begin{bmatrix} \mathbf{b}_1'^H \\ \mathbf{0}_{1\times M} \\ \vdots \\ \mathbf{0}_{1\times M} \end{bmatrix}. \tag{3.55}$$

The maximum SNR filtering matrix for the estimation of $\mathbf{x}'$ is then

$$\begin{aligned} \mathbf{H}'_{\rm max} &= \mathbf{T}'\widetilde{\mathbf{H}}'_{\rm max} \\ &= \mathbf{b}_1'\mathbf{b}_1'^H. \end{aligned} \tag{3.56}$$

Finally, the maximum SNR filtering matrix for the estimation of $\mathbf{x}$ is

$$\begin{aligned} \mathbf{H}_{\rm max} &= \mathbf{\Phi}_{\mathbf{v}}^{1/2}\mathbf{H}'_{\rm max}\mathbf{\Phi}_{\mathbf{v}}^{-1/2} \\ &= \mathbf{\Phi}_{\mathbf{v}}\mathbf{b}_1\mathbf{b}_1^H, \end{aligned} \tag{3.57}$$

which is equal to the one derived in Chapter 2.

3.4.2 Wiener

The Wiener filtering matrix is obtained by minimizing $J\left(\widetilde{\mathbf{H}}'\right)$. We get

$$\widetilde{\mathbf{H}}'_{\mathrm{W}} = \mathbf{\Lambda}_{\widetilde{\mathbf{x}}'} \mathbf{T}'^{H} \mathbf{\Phi}_{\mathbf{y}'}^{-1}. \tag{3.58}$$

We deduce that the equivalent Wiener filtering matrix for the estimation of the vector $\mathbf{x}'$ is

$$\begin{aligned} \mathbf{H}'_{\mathrm{W}} &= \mathbf{T}'\widetilde{\mathbf{H}}'_{\mathrm{W}} \\ &= \mathbf{\Phi}_{\mathbf{x}'} \mathbf{\Phi}_{\mathbf{y}'}^{-1}. \end{aligned} \tag{3.59}$$

Finally, the Wiener filtering matrix for the estimation of $\mathbf{x}$ is

$$\begin{aligned} \mathbf{H}_{\mathrm{W}} &= \mathbf{\Phi}_{\mathbf{v}}^{1/2} \mathbf{H}'_{\mathrm{W}} \mathbf{\Phi}_{\mathbf{v}}^{-1/2} \\ &= \mathbf{\Phi}_{\mathbf{x}} \mathbf{\Phi}_{\mathbf{y}}^{-1}, \end{aligned} \tag{3.60}$$

which is clearly equal to the one derived in Chapter 2.

Another insightful way to write the Wiener filter is

$$\begin{aligned} \mathbf{H}_{\mathrm{W}} &= \mathbf{\Phi}_{\mathbf{x}} \sum_{p=1}^{P} \frac{\mathbf{b}_p \mathbf{b}_p^H}{1+\lambda_p} \\ &= \mathbf{\Phi}_{\mathbf{v}} \sum_{p=1}^{P} \frac{\lambda_p}{1+\lambda_p} \mathbf{b}_p \mathbf{b}_p^H. \end{aligned} \tag{3.61}$$

From the proposed formulation, we see clearly how $\mathbf{H}_{\mathrm{W}}$ and $\mathbf{H}_{\max}$ are related. Besides a (slight) different weighting factor, $\mathbf{H}_{\mathrm{W}}$ considers all directions where speech is present, while $\mathbf{H}_{\max}$ relies only on the direction where the maximum of speech energy is present.

3.4.3 MVDR

From the criterion:

$$\min_{\widetilde{\mathbf{H}}'} \operatorname{tr}\left(\widetilde{\mathbf{H}}'\widetilde{\mathbf{H}}'^{H}\right) \quad \text{subject to } \widetilde{\mathbf{H}}'\mathbf{T}' = \mathbf{I}_P, \tag{3.62}$$

we easily find the MVDR filtering matrix:

$$\begin{aligned} \widetilde{\mathbf{H}}'_{\mathrm{MVDR}} &= \left(\mathbf{T}'^{H}\mathbf{T}'\right)^{-1} \mathbf{T}'^{H} \\ &= \mathbf{T}'^{H}. \end{aligned} \tag{3.63}$$

We deduce that the MVDR for the estimation of $\mathbf{x}'$ is

$$\begin{aligned}\mathbf{H}'_{\text{MVDR}} &= \mathbf{T}'\widetilde{\mathbf{H}}'_{\text{MVDR}} \\ &= \mathbf{T}'\mathbf{T}'^H.\end{aligned} \tag{3.64}$$

Finally, the MVDR for the estimation of $\mathbf{x}$ is

$$\mathbf{H}_{\text{MVDR}} = \mathbf{\Phi}_{\mathbf{v}} \sum_{p=1}^{P} \mathbf{b}_p \mathbf{b}_p^H. \tag{3.65}$$

For $P = M$, the MVDR filtering matrix simplifies to the identity matrix, i.e., $\mathbf{H}_{\text{MVDR}} = \mathbf{I}_M$. It is worth comparing $\mathbf{H}_{\text{MVDR}}$ with $\mathbf{H}_{\max}$ and $\mathbf{H}_{\text{W}}$.

From the obvious relationship between the MVDR and maximum SNR filtering matrices, we can deduce a whole class of minimum distortion filtering matrices:

$$\mathbf{H}_{\text{MD},P'} = \mathbf{\Phi}_{\mathbf{v}} \sum_{p'=1}^{P'} \mathbf{b}_{p'} \mathbf{b}_{p'}^H, \tag{3.66}$$

where $1 \leq P' \leq P$. We observe that $\mathbf{H}_{\text{MD},1} = \mathbf{H}_{\max}$ and $\mathbf{H}_{\text{MD},P} = \mathbf{H}_{\text{MVDR}}$. Also, we have

$$\text{oSNR}\left(\mathbf{H}_{\text{MD},P}\right) \leq \text{oSNR}\left(\mathbf{H}_{\text{MD},P-1}\right) \leq \cdots \leq \text{oSNR}\left(\mathbf{H}_{\text{MD},1}\right) = \lambda_1 \tag{3.67}$$

and

$$0 = \upsilon_{\text{sd}}\left(\mathbf{H}_{\text{MD},P}\right) \leq \upsilon_{\text{sd}}\left(\mathbf{H}_{\text{MD},P-1}\right) \leq \cdots \leq \upsilon_{\text{sd}}\left(\mathbf{H}_{\text{MD},1}\right). \tag{3.68}$$

3.4.4 Tradeoff

By minimizing the speech distortion index with the constraint that the noise reduction factor is equal to a positive value that is greater than 1, we get the tradeoff filtering matrix:

$$\widetilde{\mathbf{H}}'_{\text{T},\mu} = \mathbf{\Lambda}_{\widetilde{\mathbf{x}}'}\mathbf{T}'^H \left(\mathbf{T}'\mathbf{\Lambda}_{\widetilde{\mathbf{x}}'}\mathbf{T}'^H + \mu\mathbf{I}_M\right)^{-1}, \tag{3.69}$$

where $\mu > 0$ is a Lagrange multiplier.

Thanks to the Woodbury's identity, we find another form of the tradeoff approach:

$$\widetilde{\mathbf{H}}'_{\text{T},\mu} = \left(\mu\mathbf{\Lambda}_{\widetilde{\mathbf{x}}'}^{-1} + \mathbf{I}_M\right)^{-1} \mathbf{T}'^H. \tag{3.70}$$

We deduce that the tradeoff filtering matrices for the estimation of $\mathbf{x}'$ and $\mathbf{x}$ are, respectively,

$$\begin{aligned}\mathbf{H}'_{\mathrm{T},\mu} &= \mathbf{T}'\widetilde{\mathbf{H}}'_{\mathrm{T},\mu} \\ &= \mathbf{T}'\left(\mu\boldsymbol{\Lambda}_{\widetilde{\mathbf{x}}'}^{-1} + \mathbf{I}_M\right)^{-1}\mathbf{T}'^H \end{aligned} \tag{3.71}$$

and

$$\begin{aligned}\mathbf{H}_{\mathrm{T},\mu} &= \boldsymbol{\Phi}_{\mathbf{v}}\mathbf{T}\left(\mu\boldsymbol{\Lambda}_{\widetilde{\mathbf{x}}'}^{-1} + \mathbf{I}_M\right)^{-1}\mathbf{T}^H \\ &= \boldsymbol{\Phi}_{\mathbf{v}}\sum_{p=1}^{P}\frac{\lambda_p}{\mu+\lambda_p}\mathbf{b}_p\mathbf{b}_p^H. \end{aligned} \tag{3.72}$$

From (3.72), we observe that $\mathbf{H}_{\mathrm{T},1} = \mathbf{H}_{\mathrm{W}}$. Also, from the previous expression, we can take $\mu = 0$; in this case, we obtain the MVDR.

From all what we have seen so far, we can propose a very general noise reduction filtering matrix:

$$\mathbf{H}_{\mu,P'} = \boldsymbol{\Phi}_{\mathbf{v}}\sum_{p'=1}^{P'}\frac{\lambda_{p'}}{\mu+\lambda_{p'}}\mathbf{b}_{p'}\mathbf{b}_{p'}^H. \tag{3.73}$$

This form encompasses most known optimal filtering matrices. Indeed, it is clear that

- $\mathbf{H}_{0,1} = \mathbf{H}_{\mathrm{max}}$,
- $\mathbf{H}_{1,P} = \mathbf{H}_{\mathrm{W}}$,
- $\mathbf{H}_{0,P} = \mathbf{H}_{\mathrm{MVDR}}$,
- $\mathbf{H}_{0,P'} = \mathbf{H}_{\mathrm{MD},P'}$,
- $\mathbf{H}_{\mu,P} = \mathbf{H}_{\mathrm{T},\mu}$.

In Table 3.1, we summarize most of the optimal filtering matrices studied until now, showing clearly how they are all closely related.

Table 3.1 Optimal filtering matrices.

Maximum SNR:	$\mathbf{H}_{\max} = \mathbf{\Phi}_{\mathbf{v}} \mathbf{b}_1 \mathbf{b}_1^H$
Wiener:	$\mathbf{H}_{\mathrm{W}} = \mathbf{\Phi}_{\mathbf{v}} \sum_{p=1}^{P} \frac{\lambda_p}{1+\lambda_p} \mathbf{b}_p \mathbf{b}_p^H$
MVDR:	$\mathbf{H}_{\mathrm{MVDR}} = \mathbf{\Phi}_{\mathbf{v}} \sum_{p=1}^{P} \mathbf{b}_p \mathbf{b}_p^H$
Minimum Distortion:	$\mathbf{H}_{\mathrm{MD},P'} = \mathbf{\Phi}_{\mathbf{v}} \sum_{p'=1}^{P'} \mathbf{b}_{p'} \mathbf{b}_{p'}^H$
Tradeoff:	$\mathbf{H}_{\mathrm{T},\mu} = \mathbf{\Phi}_{\mathbf{v}} \sum_{p=1}^{P} \frac{\lambda_p}{\mu+\lambda_p} \mathbf{b}_p \mathbf{b}_p^H$
General:	$\mathbf{H}_{\mu,P'} = \mathbf{\Phi}_{\mathbf{v}} \sum_{p'=1}^{P'} \frac{\lambda_{p'}}{\mu+\lambda_{p'}} \mathbf{b}_{p'} \mathbf{b}_{p'}^H$

3.5 ANOTHER SIGNAL MODEL

Another interesting way to exploit joint diagonalization is to use the signal model:

$$\begin{aligned} \mathbf{y}' &= \mathbf{B}^H \mathbf{y} \\ &= \mathbf{B}^H \mathbf{x} + \mathbf{B}^H \mathbf{v} \\ &= \mathbf{x}' + \mathbf{v}', \end{aligned} \tag{3.74}$$

where $\mathbf{x}' = \mathbf{B}^H \mathbf{x}$ is the new desired signal vector that we want to recover from $\mathbf{y}'$. Since $\mathbf{B}$ is full rank, it is always possible to recover $\mathbf{x}$. However, all measures will be defined with respect to $\mathbf{x}'$ and $\mathbf{v}'$. The model given in (3.74) simplifies the analysis and comparison of the noise reduction filtering matrices. Indeed, we see that the correlation matrix of $\mathbf{y}'$ is diagonal:

$$\begin{aligned} \mathbf{\Phi}_{\mathbf{y}'} &= \mathbf{\Phi}_{\mathbf{x}'} + \mathbf{\Phi}_{\mathbf{v}'} \\ &= \mathbf{\Lambda} + \mathbf{I}_M, \end{aligned} \tag{3.75}$$

where $\mathbf{\Phi}_{\mathbf{x}'} = \mathbf{\Lambda}$ and $\mathbf{\Phi}_{\mathbf{v}'} = \mathbf{I}_M$ are also diagonal matrices.

The vector $\mathbf{x}'$ is estimated as follows:

$$\begin{aligned} \mathbf{z}' &= \mathbf{H}'\mathbf{y}' \\ &= \mathbf{x}'_{\mathrm{fd}} + \mathbf{v}'_{\mathrm{rn}}, \end{aligned} \tag{3.76}$$

where $\mathbf{H}'$ is a filtering matrix of size $M \times M$,

$$\mathbf{x}'_{\mathrm{fd}} = \mathbf{H}'\mathbf{x}' \tag{3.77}$$

is the filtered desired signal, and

$$\mathbf{v}'_{\text{rn}} = \mathbf{H}'\mathbf{v}' \tag{3.78}$$

is the residual noise. We deduce that the correlation matrix of $\mathbf{z}'$ is

$$\mathbf{\Phi}_{\mathbf{z}'} = \mathbf{\Phi}_{\mathbf{x}'_{\text{fd}}} + \mathbf{\Phi}_{\mathbf{v}'_{\text{rn}}}, \tag{3.79}$$

where

$$\mathbf{\Phi}_{\mathbf{x}'_{\text{fd}}} = \mathbf{H}'\mathbf{\Lambda}\mathbf{H}'^{H}, \tag{3.80}$$

$$\mathbf{\Phi}_{\mathbf{v}'_{\text{rn}}} = \mathbf{H}'\mathbf{H}'^{H}. \tag{3.81}$$

Since we consider (3.74) as the signal model for noise reduction, we find that the input SNR is

$$\text{iSNR}' = \frac{\text{tr}\left(\mathbf{\Phi}_{\mathbf{x}'}\right)}{\text{tr}\left(\mathbf{\Phi}_{\mathbf{v}'}\right)} = \frac{\text{tr}\left(\mathbf{\Lambda}\right)}{M}. \tag{3.82}$$

The output SNR is obtained from (3.79), i.e.,

$$\text{oSNR}\left(\mathbf{H}'\right) = \frac{\text{tr}\left(\mathbf{\Phi}_{\mathbf{x}'_{\text{fd}}}\right)}{\text{tr}\left(\mathbf{\Phi}_{\mathbf{v}'_{\text{rn}}}\right)} = \frac{\text{tr}\left(\mathbf{H}'\mathbf{\Lambda}\mathbf{H}'^{H}\right)}{\text{tr}\left(\mathbf{H}'\mathbf{H}'^{H}\right)} \tag{3.83}$$

and, clearly, the noise reduction factor is

$$\xi_{\text{nr}}\left(\mathbf{H}'\right) = \frac{\text{tr}\left(\mathbf{\Phi}_{\mathbf{v}'}\right)}{\text{tr}\left(\mathbf{\Phi}_{\mathbf{v}'_{\text{rn}}}\right)} = \frac{M}{\text{tr}\left(\mathbf{H}'\mathbf{H}'^{H}\right)}. \tag{3.84}$$

The speech reduction factor and the speech distortion index are, respectively,

$$\xi_{\text{sr}}\left(\mathbf{H}'\right) = \frac{\text{tr}\left(\mathbf{\Phi}_{\mathbf{x}'}\right)}{\text{tr}\left(\mathbf{\Phi}_{\mathbf{x}'_{\text{fd}}}\right)} = \frac{\text{tr}\left(\mathbf{\Lambda}\right)}{\text{tr}\left(\mathbf{H}'\mathbf{\Lambda}\mathbf{H}'^{H}\right)} \tag{3.85}$$

and

$$\upsilon_{\text{sd}}\left(\mathbf{H}'\right) = \frac{\text{tr}\left[\left(\mathbf{H}' - \mathbf{I}_M\right)\boldsymbol{\Lambda}\left(\mathbf{H}' - \mathbf{I}_M\right)^H\right]}{\text{tr}\left(\boldsymbol{\Lambda}\right)}. \tag{3.86}$$

We also have

$$\frac{\text{oSNR}\left(\mathbf{H}'\right)}{\text{iSNR}'} = \frac{\xi_{\text{nr}}\left(\mathbf{H}'\right)}{\xi_{\text{sr}}\left(\mathbf{H}'\right)}. \tag{3.87}$$

The error signal vector between the estimated and desired signals is

$$\begin{aligned}\mathbf{e}' &= \mathbf{z}' - \mathbf{x}' \\ &= \mathbf{e}'_{\text{ds}} + \mathbf{e}'_{\text{rs}},\end{aligned} \tag{3.88}$$

where

$$\mathbf{e}'_{\text{ds}} = \left(\mathbf{H} - \mathbf{I}_M\right)\mathbf{x}' \tag{3.89}$$

and

$$\mathbf{e}'_{\text{rs}} = \mathbf{H}'\mathbf{v}'. \tag{3.90}$$

The MSE criterion is then

$$\begin{aligned}J\left(\mathbf{H}'\right) &= \text{tr}\left[E\left(\mathbf{e}'\mathbf{e}'^H\right)\right] \\ &= \text{tr}\left(\boldsymbol{\Lambda}\right) + \text{tr}\left(\mathbf{H}'\boldsymbol{\Lambda}\mathbf{H}'^H\right) + \text{tr}\left(\mathbf{H}'\mathbf{H}'^H\right) - \text{tr}\left(\mathbf{H}'\boldsymbol{\Lambda}\right) - \text{tr}\left(\boldsymbol{\Lambda}\mathbf{H}'^H\right) \\ &= J_{\text{ds}}\left(\mathbf{H}'\right) + J_{\text{rs}}\left(\mathbf{H}'\right),\end{aligned} \tag{3.91}$$

where

$$J_{\text{ds}}\left(\mathbf{H}'\right) = \text{tr}\left[\left(\mathbf{H}' - \mathbf{I}_M\right)\boldsymbol{\Lambda}\left(\mathbf{H}' - \mathbf{I}_M\right)^H\right] \tag{3.92}$$

and

$$J_{\text{rs}}\left(\mathbf{H}'\right) = \text{tr}\left(\mathbf{H}'\mathbf{H}'^H\right). \tag{3.93}$$

It is easy to show that the maximum SNR filtering matrix that minimizes $J_{\text{ds}}\left(\mathbf{H}'\right)$ is

$$\mathbf{H}'_{\text{max}} = \begin{bmatrix} 1 & 0 & \cdots & 0 \\ 0 & 0 & \cdots & 0 \\ \vdots & \vdots & \ddots & \vdots \\ 0 & 0 & \cdots & 0 \end{bmatrix}. \tag{3.94}$$

It can be verified that, indeed,

$$\text{oSNR}\left(\mathbf{H}'_{\text{max}}\right) = \lambda_1 \geq \text{iSNR}', \tag{3.95}$$

and

$$\xi_{\text{nr}}\left(\mathbf{H}'_{\text{max}}\right) = M, \tag{3.96}$$

$$\xi_{\text{sr}}\left(\mathbf{H}'_{\text{max}}\right) = \frac{\text{tr}\left(\mathbf{\Lambda}\right)}{\lambda_1} \geq 1, \tag{3.97}$$

$$\upsilon_{\text{sd}}\left(\mathbf{H}'_{\text{max}}\right) = 1 - \frac{\lambda_1}{\text{tr}\left(\mathbf{\Lambda}\right)} \leq 1. \tag{3.98}$$

We deduce that the gain in SNR is

$$\frac{\text{oSNR}\left(\mathbf{H}'_{\text{max}}\right)}{\text{iSNR}'} = M\frac{\lambda_1}{\text{tr}\left(\mathbf{\Lambda}\right)} \leq M. \tag{3.99}$$

In the particular case where $\mathbf{\Phi}_{\mathbf{x}'}$ is a matrix of rank equal to 1, $\mathbf{H}'_{\text{max}}$ is distortionless [i.e., $\upsilon_{\text{sd}}\left(\mathbf{H}'_{\text{max}}\right) = 0$] and gives the maximum gain in SNR (equal to M). Eventually, the equivalent filtering matrix for the estimation of $\mathbf{x}$ is

$$\begin{aligned}\mathbf{H}_{\text{max}} &= \mathbf{B}^{-H}\mathbf{H}'_{\text{max}}\mathbf{B}^H \\ &= \mathbf{\Phi}_{\mathbf{v}}\mathbf{B}\mathbf{H}'_{\text{max}}\mathbf{B}^H \\ &= \mathbf{\Phi}_{\mathbf{v}}\mathbf{b}_1\mathbf{b}_1^H.\end{aligned} \tag{3.100}$$

The minimization of $J\left(\mathbf{H}'\right)$ leads to the Wiener filtering matrix:

$$\mathbf{H}'_{\text{W}} = \mathbf{\Lambda}\left(\mathbf{\Lambda} + \mathbf{I}_M\right)^{-1}. \tag{3.101}$$

In a more general way, it is not hard to find the tradeoff filtering matrix:

$$\mathbf{H}'_{\text{T},\mu} = \mathbf{\Lambda}\left(\mathbf{\Lambda} + \mu\mathbf{I}_M\right)^{-1}, \tag{3.102}$$

where $\mu > 0$ is a Lagrange multiplier. For $\mu = 1$, we get the Wiener filtering matrix. We have

$$\text{oSNR}\left(\mathbf{H}'_{\text{T},\mu}\right) = \frac{\text{tr}\left[\mathbf{\Lambda}^3\left(\mathbf{\Lambda} + \mu\mathbf{I}_M\right)^{-2}\right]}{\text{tr}\left[\mathbf{\Lambda}^2\left(\mathbf{\Lambda} + \mu\mathbf{I}_M\right)^{-2}\right]} \geq \text{iSNR}', \tag{3.103}$$

and

$$\xi_{\text{nr}}\left(\mathbf{H}'_{\text{T},\mu}\right) = \frac{M}{\text{tr}\left[\mathbf{\Lambda}^2\left(\mathbf{\Lambda} + \mu\mathbf{I}_M\right)^{-2}\right]} \geq 1, \tag{3.104}$$

$$\xi_{\text{sr}}\left(\mathbf{H}'_{\text{T},\mu}\right) = \frac{\text{tr}\left(\mathbf{\Lambda}\right)}{\text{tr}\left[\mathbf{\Lambda}^3\left(\mathbf{\Lambda}+\mu\mathbf{I}_M\right)^{-2}\right]} \geq 1, \tag{3.105}$$

$$\upsilon_{\text{sd}}\left(\mathbf{H}'_{\text{T},\mu}\right) = \mu^2 \frac{\text{tr}\left[\mathbf{\Lambda}\left(\mathbf{\Lambda}+\mu\mathbf{I}_M\right)^{-2}\right]}{\text{tr}\left(\mathbf{\Lambda}\right)} \leq 1. \tag{3.106}$$

We also have

$$\text{oSNR}\left(\mathbf{H}'_{\text{T},\infty}\right) = \frac{\text{tr}\left(\mathbf{\Lambda}^3\right)}{\text{tr}\left(\mathbf{\Lambda}^2\right)} \leq \lambda_1 \tag{3.107}$$

and

$$\upsilon_{\text{sd}}\left(\mathbf{H}'_{\text{T},\infty}\right) = 1. \tag{3.108}$$

The two previous equations suggest that $\mathbf{H}'_{\max}$ is more interesting than $\mathbf{H}'_{\text{T},\mu}$ for large values of μ.

Let us define the $M \times M$ matrix: $\mathbf{I}'_p$, where all its elements are equal to 0 except for its pth diagonal component which is equal to 1. It's not hard to check that the MVDR is

$$\mathbf{H}'_{\text{MVDR}} = \sum_{p=1}^{P} \mathbf{I}'_p. \tag{3.109}$$

We also find that the MD filtering matrix is

$$\mathbf{H}'_{\text{MD},P'} = \sum_{p'=1}^{P'} \mathbf{I}'_{p'}, \tag{3.110}$$

where $1 \leq P' \leq P$. We observe that $\mathbf{H}'_{\text{MD},1} = \mathbf{H}'_{\max}$ and $\mathbf{H}'_{\text{MD},P} = \mathbf{H}'_{\text{MVDR}}$. Finally, we can deduce the general filtering matrix:

$$\mathbf{H}'_{\mu,P'} = \sum_{p'=1}^{P'} \frac{\lambda_{p'}}{\mu + \lambda_{p'}} \mathbf{I}'_{p'}, \tag{3.111}$$

where $\mathbf{H}'_{1,P} = \mathbf{H}'_{\text{W}}$ and $\mathbf{H}'_{\mu,P} = \mathbf{H}'_{\text{T},\mu}$.

The previous chapter was dedicated to the study of the general speech enhancement problem as a function of the relevant parameters of the diagonalization of the speech correlation matrix. In this chapter, we roughly showed the same results (and more possibilities) but with the parameters of the joint diagonalization of the speech and noise correlation matrices. Since the two approaches are close to each other, we will stick in the rest of this work with the single diagonalization only for the sake of simplicity.

REFERENCES

[1] S.H. Jensen, P.C. Hansen, S.D. Hansen, J.A. Sørensen, Reduction of broad-band noise in speech by truncated QSVD, IEEE Trans. Speech Audio Process. 3 (1995) 439–448.

[2] Y. Hu, P.C. Loizou, A subspace approach for enhancing speech corrupted by colored noise, IEEE Signal Process. Lett. 9 (2002) 204–206.

[3] J.N. Franklin, Matrix Theory, Prentice-Hall, Englewood Cliffs, NJ, 1968.

CHAPTER 4

Single-Channel Speech Enhancement in the Time Domain

In this chapter, we study the problem of noise reduction in the time domain with a single sensor. In the first part, we assume that the speech correlation matrix is rank deficient. With this assumption, we show the importance of the speech subspace in the optimal filtering matrices. In the second part of this chapter, we revisit the problem of noise reduction without the previous assumption. This approach also leads to some interesting results which are slightly different as compared to the ones in the first part.

4.1 SIGNAL MODEL AND PROBLEM FORMULATION

The speech enhancement (or noise reduction) problem considered in this chapter is one of recovering the desired signal (or clean speech) $x(k)$, k being the discrete-time index, from the noisy observation (sensor signal) [1–3]:

$$y(k) = x(k) + v(k), \tag{4.1}$$

where $v(k)$ is the unwanted additive noise, which is assumed to be uncorrelated with $x(k)$. All signals are considered to be real, zero mean, broadband, and stationary.

Alternatively, the signal model given in (4.1) can be put into a vector form by considering the L most recent successive time samples of the noisy signal, i.e.,

$$\mathbf{y}(k) = \mathbf{x}(k) + \mathbf{v}(k), \tag{4.2}$$

where

$$\mathbf{y}(k) = \left[\, y(k)\ y(k-1)\ \cdots\ y(k-L+1) \,\right]^T \tag{4.3}$$

is a vector of length L, and $\mathbf{x}(k)$ and $\mathbf{v}(k)$ are defined in a similar way to $\mathbf{y}(k)$ from (4.3). Since $x(k)$ and $v(k)$ are uncorrelated by assumption, the correlation matrix (of size $L \times L$) of the noisy signal can be written as

$$\begin{aligned} \mathbf{R}_{\mathbf{y}} &= E\left[\mathbf{y}(k)\mathbf{y}^T(k)\right] \\ &= \mathbf{R}_{\mathbf{x}} + \mathbf{R}_{\mathbf{v}}, \end{aligned} \tag{4.4}$$

Speech Enhancement. http://dx.doi.org/10.1016/B978-0-12-800139-4.00004-9

where $\mathbf{R_x} = E\left[\mathbf{x}(k)\mathbf{x}^T(k)\right]$ and $\mathbf{R_v} = E\left[\mathbf{v}(k)\mathbf{v}^T(k)\right]$ are the correlation matrices of $\mathbf{x}(k)$ and $\mathbf{v}(k)$, respectively. The noise correlation matrix, $\mathbf{R_v}$, is assumed to be full rank, i.e., equal to L. We assume for now that the rank of the desired signal correlation matrix, $\mathbf{R_x}$, is equal to P, where P is smaller than L. This assumption is often assumed reasonable in the literature since the speech signal can be roughly modeled as the sum of a small number of sinusoids (equal to P). In any case, we can always choose L much greater than P. Then, the objective of single-channel speech enhancement (or noise reduction) is to estimate the desired signal vector, $\mathbf{x}(k)$, or any known linear transformation of it from $\mathbf{y}(k)$. This should be done in such a way that the noise is reduced as much as possible with no or little distortion of the desired signal.

As it was explained in Chapter 2, the desired signal correlation matrix can be diagonalized as

$$\mathbf{Q}_\mathbf{x}^T \mathbf{R_x} \mathbf{Q_x} = \mathbf{\Lambda_x}, \tag{4.5}$$

where

$$\mathbf{Q_x} = \left[\mathbf{q}_{\mathbf{x},1}\ \mathbf{q}_{\mathbf{x},2}\ \cdots\ \mathbf{q}_{\mathbf{x},L}\right] \tag{4.6}$$

is an orthogonal matrix, i.e., $\mathbf{Q}_\mathbf{x}^T\mathbf{Q_x} = \mathbf{Q_x}\mathbf{Q}_\mathbf{x}^T = \mathbf{I}_L$, with $\mathbf{I}_L$ being the $L \times L$ identity matrix, and

$$\mathbf{\Lambda_x} = \text{diag}\left(\lambda_{\mathbf{x},1}, \lambda_{\mathbf{x},2}, \ldots, \lambda_{\mathbf{x},L}\right) \tag{4.7}$$

is a diagonal matrix, with $\lambda_{\mathbf{x},1} \geq \lambda_{\mathbf{x},2} \geq \cdots \geq \lambda_{\mathbf{x},P} > 0$ and $\lambda_{\mathbf{x},P+1} = \lambda_{\mathbf{x},P+2} = \cdots = \lambda_{\mathbf{x},L} = 0$. Let

$$\mathbf{Q_x} = \left[\mathbf{T_x}\ \mathbf{\Xi_x}\right], \tag{4.8}$$

where the $L \times P$ matrix $\mathbf{T_x}$ contains the eigenvectors corresponding to the nonzero eigenvalues of $\mathbf{R_x}$ and the $L \times (L - P)$ matrix $\mathbf{\Xi_x}$ contains the eigenvectors corresponding to the null eigenvalues of $\mathbf{R_x}$. The desired signal vector can then be expressed as

$$\begin{aligned} \mathbf{x}(k) &= \mathbf{Q_x}\mathbf{Q}_\mathbf{x}^T\mathbf{x}(k) \\ &= \mathbf{T_x}\mathbf{T}_\mathbf{x}^T\mathbf{x}(k) \\ &= \mathbf{T_x}\widetilde{\mathbf{x}}(k), \end{aligned} \tag{4.9}$$

where

$$\widetilde{\mathbf{x}}(k) = \mathbf{T}_\mathbf{x}^T\mathbf{x}(k) \tag{4.10}$$

is the transformed desired signal vector of length P. Therefore, the signal model for noise reduction becomes

$$\mathbf{y}(k) = \mathbf{T}_{\mathbf{x}}\widetilde{\mathbf{x}}(k) + \mathbf{v}(k), \tag{4.11}$$

for which its correlation matrix is

$$\begin{aligned}\mathbf{R}_{\mathbf{y}} &= \mathbf{T}_{\mathbf{x}}\mathbf{R}_{\widetilde{\mathbf{x}}}\mathbf{T}_{\mathbf{x}}^T + \mathbf{R}_{\mathbf{v}} \\ &= \mathbf{T}_{\mathbf{x}}\mathbf{\Lambda}_{\widetilde{\mathbf{x}}}\mathbf{T}_{\mathbf{x}}^T + \mathbf{R}_{\mathbf{v}},\end{aligned} \tag{4.12}$$

where

$$\begin{aligned}\mathbf{R}_{\widetilde{\mathbf{x}}} &= E\left[\widetilde{\mathbf{x}}(k)\widetilde{\mathbf{x}}^T(k)\right] \\ &= \mathbf{T}_{\mathbf{x}}^T\mathbf{R}_{\mathbf{x}}\mathbf{T}_{\mathbf{x}} \\ &= \mathbf{T}_{\mathbf{x}}^T\mathbf{Q}_{\mathbf{x}}\mathbf{\Lambda}_{\mathbf{x}}\mathbf{Q}_{\mathbf{x}}^T\mathbf{T}_{\mathbf{x}} \\ &= \text{diag}\left(\lambda_{\mathbf{x},1}, \lambda_{\mathbf{x},2}, \ldots, \lambda_{\mathbf{x},P}\right) \\ &= \mathbf{\Lambda}_{\widetilde{\mathbf{x}}}\end{aligned} \tag{4.13}$$

and, obviously, $\mathbf{R}_{\mathbf{x}} = \mathbf{T}_{\mathbf{x}}\mathbf{R}_{\widetilde{\mathbf{x}}}\mathbf{T}_{\mathbf{x}}^T = \mathbf{T}_{\mathbf{x}}\mathbf{\Lambda}_{\widetilde{\mathbf{x}}}\mathbf{T}_{\mathbf{x}}^T$.

4.2 LINEAR FILTERING WITH A RECTANGULAR MATRIX

From the general linear filtering approach [1,3–6], we can estimate the transformed desired signal vector, $\widetilde{\mathbf{x}}(k)$, by applying a linear transformation to the observation signal vector, $\mathbf{y}(k)$, i.e.,

$$\begin{aligned}\widetilde{\mathbf{z}}(k) &= \widetilde{\mathbf{H}}\mathbf{y}(k) \\ &= \widetilde{\mathbf{H}}\left[\mathbf{x}(k) + \mathbf{v}(k)\right] \\ &= \widetilde{\mathbf{x}}_{\text{fd}}(k) + \widetilde{\mathbf{v}}_{\text{rn}}(k),\end{aligned} \tag{4.14}$$

where $\widetilde{\mathbf{z}}(k)$ is supposed to be the estimate of $\widetilde{\mathbf{x}}(k)$,

$$\widetilde{\mathbf{H}} = \begin{bmatrix} \widetilde{\mathbf{h}}_1^T \\ \widetilde{\mathbf{h}}_2^T \\ \vdots \\ \widetilde{\mathbf{h}}_P^T \end{bmatrix} \tag{4.15}$$

is a rectangular filtering matrix of size $P \times L$,

$$\widetilde{\mathbf{h}}_p = \left[\widetilde{h}_{p,0}\ \widetilde{h}_{p,1} \cdots \widetilde{h}_{p,L-1}\right]^T, \quad p = 1, 2, \ldots, P \tag{4.16}$$

are real-valued filters of length L,

$$\begin{aligned}\widetilde{\mathbf{x}}_{\text{fd}}(k) &= \widetilde{\mathbf{H}}\mathbf{x}(k)\\ &= \widetilde{\mathbf{H}}\mathbf{T}_{\mathbf{x}}\widetilde{\mathbf{x}}(k)\end{aligned} \tag{4.17}$$

is the filtered transformed desired signal, and

$$\widetilde{\mathbf{v}}_{\text{rn}}(k) = \widetilde{\mathbf{H}}\mathbf{v}(k) \tag{4.18}$$

is the residual noise. As a result, the estimate of $\mathbf{x}(k)$ is supposed to be

$$\begin{aligned}\mathbf{z}(k) &= \mathbf{T}_{\mathbf{x}}\widetilde{\mathbf{z}}(k)\\ &= \mathbf{T}_{\mathbf{x}}\widetilde{\mathbf{H}}\mathbf{y}(k)\\ &= \mathbf{H}\mathbf{y}(k),\end{aligned} \tag{4.19}$$

where

$$\mathbf{H} = \mathbf{T}_{\mathbf{x}}\widetilde{\mathbf{H}} = \begin{bmatrix}\mathbf{h}_1^T\\ \mathbf{h}_2^T\\ \vdots\\ \mathbf{h}_L^T\end{bmatrix} \tag{4.20}$$

is the filtering matrix of size $L \times L$ leading to the estimation of $\mathbf{x}(k)$. The correlation matrix of $\widetilde{\mathbf{z}}(k)$ is then

$$\begin{aligned}\mathbf{R}_{\widetilde{\mathbf{z}}} &= E\left[\widetilde{\mathbf{z}}(k)\widetilde{\mathbf{z}}^T(k)\right]\\ &= \mathbf{R}_{\widetilde{\mathbf{x}}_{\text{fd}}} + \mathbf{R}_{\widetilde{\mathbf{v}}_{\text{rn}}},\end{aligned} \tag{4.21}$$

where

$$\begin{aligned}\mathbf{R}_{\widetilde{\mathbf{x}}_{\text{fd}}} &= \widetilde{\mathbf{H}}\mathbf{R}_{\mathbf{x}}\widetilde{\mathbf{H}}^T\\ &= \widetilde{\mathbf{H}}\mathbf{T}_{\mathbf{x}}\mathbf{\Lambda}_{\widetilde{\mathbf{x}}}\mathbf{T}_{\mathbf{x}}^T\widetilde{\mathbf{H}}^T,\end{aligned} \tag{4.22}$$

$$\mathbf{R}_{\widetilde{\mathbf{v}}_{\text{rn}}} = \widetilde{\mathbf{H}}\mathbf{R}_{\mathbf{v}}\widetilde{\mathbf{H}}^T. \tag{4.23}$$

We also observe that $\mathbf{R}_{\mathbf{z}} = \mathbf{T}_{\mathbf{x}}\mathbf{R}_{\widetilde{\mathbf{z}}}\mathbf{T}_{\mathbf{x}}^T$ and $\text{tr}\left(\mathbf{R}_{\mathbf{z}}\right) = \text{tr}\left(\mathbf{R}_{\widetilde{\mathbf{z}}}\right)$.

4.3 PERFORMANCE MEASURES

The performance measures considered here are very similar to the ones derived in Chapter 2. Therefore, in this section, we only briefly expose the most useful ones for the evaluation of noise reduction and speech distortion of the filtering matrices.

The input and output SNRs are defined, respectively, as

$$\begin{aligned}\mathrm{iSNR} &= \frac{\mathrm{tr}\left(\mathbf{R}_{\mathbf{x}}\right)}{\mathrm{tr}\left(\mathbf{R}_{\mathbf{v}}\right)} \\ &= \frac{\sigma_x^2}{\sigma_v^2}\end{aligned} \tag{4.24}$$

and

$$\begin{aligned}\mathrm{oSNR}\left(\widetilde{\mathbf{H}}\right) &= \frac{\mathrm{tr}\left(\mathbf{R}_{\widetilde{\mathbf{x}}_{\mathrm{fd}}}\right)}{\mathrm{tr}\left(\mathbf{R}_{\widetilde{\mathbf{v}}_{\mathrm{rn}}}\right)} \\ &= \frac{\mathrm{tr}\left(\widetilde{\mathbf{H}}\mathbf{R}_{\mathbf{x}}\widetilde{\mathbf{H}}^T\right)}{\mathrm{tr}\left(\widetilde{\mathbf{H}}\mathbf{R}_{\mathbf{v}}\widetilde{\mathbf{H}}^T\right)},\end{aligned} \tag{4.25}$$

where $\sigma_x^2 = E\left[x^2(k)\right]$ and $\sigma_v^2 = E\left[v^2(k)\right]$ are the variances of $x(k)$ and $v(k)$, respectively. The noise reduction factor is

$$\xi_{\mathrm{nr}}\left(\widetilde{\mathbf{H}}\right) = \frac{\mathrm{tr}\left(\mathbf{R}_{\mathbf{v}}\right)}{\mathrm{tr}\left(\widetilde{\mathbf{H}}\mathbf{R}_{\mathbf{v}}\widetilde{\mathbf{H}}^T\right)}. \tag{4.26}$$

The speech reduction factor is given by

$$\begin{aligned}\xi_{\mathrm{sr}}\left(\widetilde{\mathbf{H}}\right) &= \frac{\mathrm{tr}\left(\mathbf{R}_{\mathbf{x}}\right)}{\mathrm{tr}\left(\mathbf{R}_{\widetilde{\mathbf{x}}_{\mathrm{fd}}}\right)} \\ &= \frac{\mathrm{tr}\left(\mathbf{\Lambda}_{\widetilde{\mathbf{x}}}\right)}{\mathrm{tr}\left(\widetilde{\mathbf{H}}\mathbf{T}_{\mathbf{x}}\mathbf{\Lambda}_{\widetilde{\mathbf{x}}}\mathbf{T}_{\mathbf{x}}^T\widetilde{\mathbf{H}}^T\right)}\end{aligned} \tag{4.27}$$

and a rectangular filtering matrix that does not affect the transformed desired signal requires the constraint:

$$\widetilde{\mathbf{H}}\mathbf{T}_{\mathbf{x}} = \mathbf{I}_P. \tag{4.28}$$

It is easy to check that

$$\frac{\mathrm{oSNR}\left(\widetilde{\mathbf{H}}\right)}{\mathrm{iSNR}} = \frac{\xi_{\mathrm{nr}}\left(\widetilde{\mathbf{H}}\right)}{\xi_{\mathrm{sr}}\left(\widetilde{\mathbf{H}}\right)}. \tag{4.29}$$

The speech distortion index is defined as

$$\upsilon_{\text{sd}}\left(\widetilde{\mathbf{H}}\right) = \frac{E\left\{\left[\widetilde{\mathbf{x}}_{\text{fd}}(k) - \widetilde{\mathbf{x}}(k)\right]^T\left[\widetilde{\mathbf{x}}_{\text{fd}}(k) - \widetilde{\mathbf{x}}(k)\right]\right\}}{\text{tr}\left(\mathbf{R}_{\widetilde{\mathbf{x}}}\right)} = \frac{\text{tr}\left[\left(\widetilde{\mathbf{H}}\mathbf{T}_{\mathbf{x}} - \mathbf{I}_P\right)\mathbf{\Lambda}_{\widetilde{\mathbf{x}}}\left(\widetilde{\mathbf{H}}\mathbf{T}_{\mathbf{x}} - \mathbf{I}_P\right)^T\right]}{\text{tr}\left(\mathbf{\Lambda}_{\widetilde{\mathbf{x}}}\right)}. \tag{4.30}$$

We define the error signal vector between the estimated and desired signals as

$$\widetilde{\mathbf{e}}(k) = \widetilde{\mathbf{z}}(k) - \widetilde{\mathbf{x}}(k) = \widetilde{\mathbf{e}}_{\text{ds}}(k) + \widetilde{\mathbf{e}}_{\text{rs}}(k), \tag{4.31}$$

where

$$\widetilde{\mathbf{e}}_{\text{ds}}(k) = \widetilde{\mathbf{x}}_{\text{fd}}(k) - \widetilde{\mathbf{x}}(k) = \left(\widetilde{\mathbf{H}}\mathbf{T}_{\mathbf{x}} - \mathbf{I}_P\right)\widetilde{\mathbf{x}}(k) \tag{4.32}$$

is the signal distortion due to the rectangular filtering matrix and

$$\widetilde{\mathbf{e}}_{\text{rs}}(k) = \widetilde{\mathbf{v}}_{\text{rn}}(k) = \widetilde{\mathbf{H}}\mathbf{v}(k) \tag{4.33}$$

represents the residual noise. The MSE criterion is then

$$J\left(\widetilde{\mathbf{H}}\right) = \text{tr}\left\{E\left[\widetilde{\mathbf{e}}(k)\widetilde{\mathbf{e}}^T(k)\right]\right\} = \text{tr}\left(\mathbf{\Lambda}_{\widetilde{\mathbf{x}}}\right) + \text{tr}\left(\widetilde{\mathbf{H}}\mathbf{R}_{\mathbf{y}}\widetilde{\mathbf{H}}^T\right) - 2\text{tr}\left(\widetilde{\mathbf{H}}\mathbf{T}_{\mathbf{x}}\mathbf{\Lambda}_{\widetilde{\mathbf{x}}}\right) = J_{\text{ds}}\left(\widetilde{\mathbf{H}}\right) + J_{\text{rs}}\left(\widetilde{\mathbf{H}}\right), \tag{4.34}$$

where

$$J_{\text{ds}}\left(\widetilde{\mathbf{H}}\right) = \text{tr}\left\{E\left[\widetilde{\mathbf{e}}_{\text{ds}}(k)\widetilde{\mathbf{e}}_{\text{ds}}^T(k)\right]\right\} = \text{tr}\left[\left(\widetilde{\mathbf{H}}\mathbf{T}_{\mathbf{x}} - \mathbf{I}_P\right)\mathbf{\Lambda}_{\widetilde{\mathbf{x}}}\left(\widetilde{\mathbf{H}}\mathbf{T}_{\mathbf{x}} - \mathbf{I}_P\right)^T\right] = \text{tr}\left(\mathbf{\Lambda}_{\widetilde{\mathbf{x}}}\right)\upsilon_{\text{sd}}\left(\widetilde{\mathbf{H}}\right) \tag{4.35}$$

and

$$
\begin{aligned}
J_{\rm rs}\left(\widetilde{\mathbf{H}}\right) &= \mathrm{tr}\left\{E\left[\widetilde{\mathbf{e}}_{\rm rs}(k)\widetilde{\mathbf{e}}_{\rm rs}^T(k)\right]\right\} \\
&= \mathrm{tr}\left(\widetilde{\mathbf{H}}\mathbf{R}_{\mathbf{v}}\widetilde{\mathbf{H}}^T\right) \\
&= \frac{\mathrm{tr}\left(\mathbf{R}_{\mathbf{v}}\right)}{\xi_{\rm nr}\left(\widetilde{\mathbf{H}}\right)}.
\end{aligned}
\tag{4.36}
$$

We deduce that

$$
\begin{aligned}
\frac{J_{\rm ds}\left(\widetilde{\mathbf{H}}\right)}{J_{\rm rs}\left(\widetilde{\mathbf{H}}\right)} &= \mathrm{iSNR} \times \xi_{\rm nr}\left(\widetilde{\mathbf{H}}\right) \times \upsilon_{\rm sd}\left(\widetilde{\mathbf{H}}\right) \\
&= \mathrm{oSNR}\left(\widetilde{\mathbf{H}}\right) \times \xi_{\rm sr}\left(\widetilde{\mathbf{H}}\right) \times \upsilon_{\rm sd}\left(\widetilde{\mathbf{H}}\right).
\end{aligned}
\tag{4.37}
$$

We observe how the MSEs are related to the performance measures.

4.4 OPTIMAL RECTANGULAR FILTERING MATRICES

In this section, we only briefly present the most important rectangular filtering matrices for noise reduction since they are similar to the ones derived in Chapter 2.

The maximum SNR filtering matrix is given by

$$
\widetilde{\mathbf{H}}_{\max} = \begin{bmatrix} \varsigma_1 \mathbf{b}_1^T \\ \varsigma_2 \mathbf{b}_1^T \\ \vdots \\ \varsigma_P \mathbf{b}_1^T \end{bmatrix},
\tag{4.38}
$$

where $\mathbf{b}_1$ is the eigenvector corresponding to the maximum eigenvalue, λ_1, of the matrix $\mathbf{R}_{\mathbf{v}}^{-1}\mathbf{R}_{\mathbf{x}}$ and

$$
\begin{aligned}
\varsigma_p &= \frac{\lambda_{\mathbf{x},p}\mathbf{b}_1^T\mathbf{q}_{\mathbf{x},p}}{\mathbf{b}_1^T\mathbf{R}_{\mathbf{x}}\mathbf{b}_1} \\
&= \frac{\lambda_{\mathbf{x},p}\mathbf{b}_1^T\mathbf{q}_{\mathbf{x},p}}{\lambda_1}, \quad p = 1, 2, \ldots, P.
\end{aligned}
\tag{4.39}
$$

We can rewrite (4.38) as

$$
\widetilde{\mathbf{H}}_{\max} = \mathbf{\Lambda}_{\widetilde{\mathbf{x}}}\mathbf{T}_{\mathbf{x}}^T\frac{\mathbf{b}_1\mathbf{b}_1^T}{\lambda_1}.
\tag{4.40}
$$

As a result,

$$\begin{aligned}\mathbf{H}_{\max} &= \mathbf{R}_{\mathbf{x}}\frac{\mathbf{b}_1\mathbf{b}_1^T}{\lambda_1} \\ &= \mathbf{R}_{\mathbf{v}}\mathbf{b}_1\mathbf{b}_1^T.\end{aligned} \tag{4.41}$$

We find that the Wiener filtering matrix is

$$\begin{aligned}\widetilde{\mathbf{H}}_{\mathrm{W}} &= \mathbf{R}_{\widetilde{\mathbf{x}}}\mathbf{T}_{\mathbf{x}}^T\mathbf{R}_{\mathbf{y}}^{-1} \\ &= \mathbf{T}_{\mathbf{x}}^T\mathbf{R}_{\mathbf{x}}\mathbf{R}_{\mathbf{y}}^{-1} \\ &= \mathbf{T}_{\mathbf{x}}^T\left(\mathbf{I}_L - \mathbf{R}_{\mathbf{v}}\mathbf{R}_{\mathbf{y}}^{-1}\right).\end{aligned} \tag{4.42}$$

Another way to express $\widetilde{\mathbf{H}}_{\mathrm{W}}$ is

$$\begin{aligned}\widetilde{\mathbf{H}}_{\mathrm{W}} &= \left(\mathbf{I}_P + \mathbf{\Lambda}_{\widetilde{\mathbf{x}}}\mathbf{T}_{\mathbf{x}}^T\mathbf{R}_{\mathbf{v}}^{-1}\mathbf{T}_{\mathbf{x}}\right)^{-1}\mathbf{\Lambda}_{\widetilde{\mathbf{x}}}\mathbf{T}_{\mathbf{x}}^T\mathbf{R}_{\mathbf{v}}^{-1} \\ &= \left(\mathbf{\Lambda}_{\widetilde{\mathbf{x}}}^{-1} + \mathbf{T}_{\mathbf{x}}^T\mathbf{R}_{\mathbf{v}}^{-1}\mathbf{T}_{\mathbf{x}}\right)^{-1}\mathbf{T}_{\mathbf{x}}^T\mathbf{R}_{\mathbf{v}}^{-1}.\end{aligned} \tag{4.43}$$

We also deduce that

$$\begin{aligned}\mathbf{H}_{\mathrm{W}} &= \mathbf{I}_L - \mathbf{R}_{\mathbf{v}}\mathbf{R}_{\mathbf{y}}^{-1} \\ &= \mathbf{T}_{\mathbf{x}}\left(\mathbf{I}_P + \mathbf{\Lambda}_{\widetilde{\mathbf{x}}}\mathbf{T}_{\mathbf{x}}^T\mathbf{R}_{\mathbf{v}}^{-1}\mathbf{T}_{\mathbf{x}}\right)^{-1}\mathbf{\Lambda}_{\widetilde{\mathbf{x}}}\mathbf{T}_{\mathbf{x}}^T\mathbf{R}_{\mathbf{v}}^{-1}\end{aligned} \tag{4.44}$$

and, if the noise is white, the previous expression simplifies to

$$\mathbf{H}_{\mathrm{W}} = \mathbf{T}_{\mathbf{x}}\left(\sigma_v^2\mathbf{I}_P + \mathbf{\Lambda}_{\widetilde{\mathbf{x}}}\right)^{-1}\mathbf{\Lambda}_{\widetilde{\mathbf{x}}}\mathbf{T}_{\mathbf{x}}^T. \tag{4.45}$$

The MVDR filtering matrix is

$$\begin{aligned}\widetilde{\mathbf{H}}_{\mathrm{MVDR}} &= \left(\mathbf{T}_{\mathbf{x}}^T\mathbf{R}_{\mathbf{v}}^{-1}\mathbf{T}_{\mathbf{x}}\right)^{-1}\mathbf{T}_{\mathbf{x}}^T\mathbf{R}_{\mathbf{v}}^{-1} \\ &= \left(\mathbf{T}_{\mathbf{x}}^T\mathbf{R}_{\mathbf{y}}^{-1}\mathbf{T}_{\mathbf{x}}\right)^{-1}\mathbf{T}_{\mathbf{x}}^T\mathbf{R}_{\mathbf{y}}^{-1}\end{aligned} \tag{4.46}$$

for the estimation of $\widetilde{\mathbf{x}}(k)$ and

$$\begin{aligned}\mathbf{H}_{\mathrm{MVDR}} &= \mathbf{T}_{\mathbf{x}}\left(\mathbf{T}_{\mathbf{x}}^T\mathbf{R}_{\mathbf{v}}^{-1}\mathbf{T}_{\mathbf{x}}\right)^{-1}\mathbf{T}_{\mathbf{x}}^T\mathbf{R}_{\mathbf{v}}^{-1} \\ &= \mathbf{T}_{\mathbf{x}}\left(\mathbf{T}_{\mathbf{x}}^T\mathbf{R}_{\mathbf{y}}^{-1}\mathbf{T}_{\mathbf{x}}\right)^{-1}\mathbf{T}_{\mathbf{x}}^T\mathbf{R}_{\mathbf{y}}^{-1}\end{aligned} \tag{4.47}$$

for the estimation of $\mathbf{x}(k)$. For $P = L$, the MVDR filtering matrix simplifies to the identity matrix, i.e., $\mathbf{H}_{\text{MVDR}} = \mathbf{I}_L$. Therefore, the higher is the dimension of the nullspace of $\mathbf{R}_\mathbf{x}$, the more the MVDR is efficient in terms of noise reduction. The best scenario corresponds to $P = 1$. If the noise is white, the MVDR simplifies to [4, 7]

$$\mathbf{H}_{\text{MVDR}} = \mathbf{T}_\mathbf{x}\mathbf{T}_\mathbf{x}^T. \tag{4.48}$$

We can easily deduce the tradeoff filtering matrix:

$$\widetilde{\mathbf{H}}_{\text{T},\mu} = \mathbf{\Lambda}_{\widetilde{\mathbf{x}}}\mathbf{T}_\mathbf{x}^T\left(\mathbf{T}_\mathbf{x}\mathbf{\Lambda}_{\widetilde{\mathbf{x}}}\mathbf{T}_\mathbf{x}^T + \mu\mathbf{R}_\mathbf{v}\right)^{-1}, \tag{4.49}$$

which can be rewritten, thanks to the Woodbury's identity, as

$$\widetilde{\mathbf{H}}_{\text{T},\mu} = \left(\mu\mathbf{\Lambda}_{\widetilde{\mathbf{x}}}^{-1} + \mathbf{T}_\mathbf{x}^T\mathbf{R}_\mathbf{v}^{-1}\mathbf{T}_\mathbf{x}\right)^{-1}\mathbf{T}_\mathbf{x}^T\mathbf{R}_\mathbf{v}^{-1}, \tag{4.50}$$

where $\mu > 0$ in (4.49) and $\mu \geq 0$ in (4.50). As a consequence,

$$\mathbf{H}_{\text{T},\mu} = \mathbf{T}_\mathbf{x}\left(\mu\mathbf{\Lambda}_{\widetilde{\mathbf{x}}}^{-1} + \mathbf{T}_\mathbf{x}^T\mathbf{R}_\mathbf{v}^{-1}\mathbf{T}_\mathbf{x}\right)^{-1}\mathbf{T}_\mathbf{x}^T\mathbf{R}_\mathbf{v}^{-1}. \tag{4.51}$$

Now, let us decompose the noise correlation matrix as

$$\mathbf{R}_\mathbf{v} = \mathbf{Q}_\mathbf{v}\mathbf{\Lambda}_\mathbf{v}\mathbf{Q}_\mathbf{v}^T, \tag{4.52}$$

where the orthogonal and diagonal matrices $\mathbf{Q}_\mathbf{v}$ and $\mathbf{\Lambda}_\mathbf{v}$ are defined similarly to $\mathbf{Q}_\mathbf{x}$ and $\mathbf{\Lambda}_\mathbf{x}$, respectively. We assume that the (positive) eigenvalues of $\mathbf{R}_\mathbf{v}$ have the following structure: $\lambda_{\mathbf{v},1} \geq \lambda_{\mathbf{v},2} \geq \cdots \geq \lambda_{\mathbf{v},Q} > \sigma_{\text{wn}}^2$ and $\lambda_{\mathbf{v},Q+1} = \lambda_{\mathbf{v},Q+2} = \cdots = \lambda_{\mathbf{v},L} = \sigma_{\text{wn}}^2$, where $P + Q \leq L$. We can rewrite $\mathbf{Q}_\mathbf{v}$ as

$$\mathbf{Q}_\mathbf{v} = \left[\mathbf{T}_\mathbf{v}\ \mathbf{\Xi}_\mathbf{v}\right], \tag{4.53}$$

where the $L \times Q$ matrix $\mathbf{T}_\mathbf{v}$ contains the eigenvectors corresponding to the first Q eigenvalues of $\mathbf{R}_\mathbf{v}$ and the $L \times (L - Q)$ matrix $\mathbf{\Xi}_\mathbf{v}$ contains the eigenvectors corresponding to the last $L - Q$ eigenvalues of $\mathbf{R}_\mathbf{v}$. As a result, the noise signal vector can be decomposed as

$$\mathbf{v}(k) = \mathbf{v}_\text{c}(k) + \mathbf{v}_\text{u}(k), \tag{4.54}$$

where

$$\mathbf{v}_\text{c}(k) = \mathbf{T}_\mathbf{v}\mathbf{T}_\mathbf{v}^T\mathbf{v}(k) \tag{4.55}$$

corresponds to the correlated noise,

$$\mathbf{v}_{\mathrm{u}}(k) = \boldsymbol{\Xi}_{\mathbf{v}}\boldsymbol{\Xi}_{\mathbf{v}}^T\mathbf{v}(k) \tag{4.56}$$

corresponds to the uncorrelated noise, and $E\left[\mathbf{v}_{\mathrm{c}}(k)\mathbf{v}_{\mathrm{u}}^T(k)\right] = \mathbf{0}_{L\times L}$. Using the constraints:

$$\widetilde{\mathbf{H}}\mathbf{C}_{\mathbf{xv}} = \left[\mathbf{I}_P\ \mathbf{0}_{P\times Q}\right], \tag{4.57}$$

where

$$\mathbf{C}_{\mathbf{xv}} = \left[\mathbf{T}_{\mathbf{x}}\ \mathbf{T}_{\mathbf{v}}\right] \tag{4.58}$$

is the constraint matrix of size $L \times (P + Q)$, we deduce that the LCMV filtering matrix is

$$\widetilde{\mathbf{H}}_{\mathrm{LCMV}} = \left[\mathbf{I}_P\ \mathbf{0}_{P\times Q}\right]\left(\mathbf{C}_{\mathbf{xv}}^T\mathbf{R}_{\mathbf{y}}^{-1}\mathbf{C}_{\mathbf{xv}}\right)^{-1}\mathbf{C}_{\mathbf{xv}}^T\mathbf{R}_{\mathbf{y}}^{-1} \tag{4.59}$$

or

$$\mathbf{H}_{\mathrm{LCMV}} = \mathbf{T}_{\mathbf{x}}\left[\mathbf{I}_P\ \mathbf{0}_{P\times Q}\right]\left(\mathbf{C}_{\mathbf{xv}}^T\mathbf{R}_{\mathbf{y}}^{-1}\mathbf{C}_{\mathbf{xv}}\right)^{-1}\mathbf{C}_{\mathbf{xv}}^T\mathbf{R}_{\mathbf{y}}^{-1}. \tag{4.60}$$

4.5 SINGLE-CHANNEL NOISE REDUCTION REVISITED

In the previous sections, it was assumed that the rank of $\mathbf{R}_{\mathbf{x}}$ is smaller than L; this, however, rarely happens in practice even though $\mathbf{R}_{\mathbf{x}}$ is usually a very ill-conditioned matrix. In this section, we show how to estimate the first P samples of the desired signal vector, $\mathbf{x}(k)$, with a subspace approach. The derived filtering matrices for noise reduction are subtly different from those of Section 4.4.

4.5.1 Orthogonal Decomposition

Let us define a short version of $\mathbf{x}(k)$:

$$\mathbf{x}_{\mathrm{s}}(k) = \left[x(k)\ x(k-1)\ \cdots\ x(k-P+1)\right]^T. \tag{4.61}$$

This vector of length P is now considered as the desired signal vector. The vector $\mathbf{x}(k)$ can be decomposed into two orthogonal components [3]:

$$\begin{aligned}\mathbf{x}(k) &= \mathbf{R}_{\mathbf{x}\mathbf{x}_{\mathrm{s}}}\mathbf{R}_{\mathbf{x}_{\mathrm{s}}}^{-1}\mathbf{x}_{\mathrm{s}}(k) + \mathbf{x}_{\mathrm{i}}(k)\\ &= \boldsymbol{\Gamma}_{\mathbf{x}\mathbf{x}_{\mathrm{s}}}\mathbf{x}_{\mathrm{s}}(k) + \mathbf{x}_{\mathrm{i}}(k)\\ &= \mathbf{x}'(k) + \mathbf{x}_{\mathrm{i}}(k),\end{aligned} \tag{4.62}$$

where

$$\mathbf{\Gamma}_{\mathbf{x}\mathbf{x}_s} = \mathbf{R}_{\mathbf{x}\mathbf{x}_s}\mathbf{R}_{\mathbf{x}_s}^{-1} \tag{4.63}$$

is the time-domain steering matrix, $\mathbf{R}_{\mathbf{x}\mathbf{x}_s} = E\left[\mathbf{x}(k)\mathbf{x}_s^T(k)\right]$ is the cross-correlation matrix of size $L \times P$ between $\mathbf{x}(k)$ and $\mathbf{x}_s(k)$, $\mathbf{R}_{\mathbf{x}_s} = E\left[\mathbf{x}_s(k)\mathbf{x}_s^T(k)\right]$ is the correlation matrix of $\mathbf{x}_s(k)$ (of size $P \times P$), $\mathbf{x}'(k) = \mathbf{\Gamma}_{\mathbf{x}\mathbf{x}_s}\mathbf{x}_s(k)$, and $\mathbf{x}_i(k)$ is what we consider as the interference signal vector, since all the information we want is in $\mathbf{x}'(k)$. Obviously, the first P components of $\mathbf{x}'(k)$ are equal to $\mathbf{x}_s(k)$, while the first P components of $\mathbf{x}_i(k)$ are equal to $\mathbf{0}_{P\times 1}$. It can be verified that $\mathbf{x}'(k)$ and $\mathbf{x}_i(k)$ are orthogonal, i.e.,

$$E\left[\mathbf{x}'(k)\mathbf{x}_i^T(k)\right] = \mathbf{0}_{L\times L}. \tag{4.64}$$

The signal model for noise reduction is now

$$\mathbf{y}(k) = \mathbf{x}'(k) + \mathbf{x}_i(k) + \mathbf{v}(k), \tag{4.65}$$

for which its correlation matrix is

$$\begin{aligned} \mathbf{R}_\mathbf{y} &= \mathbf{R}_{\mathbf{x}'} + \mathbf{R}_{\mathbf{x}_i} + \mathbf{R}_\mathbf{v} \\ &= \mathbf{R}_{\mathbf{x}'} + \mathbf{R}_{\text{in}}, \end{aligned} \tag{4.66}$$

where $\mathbf{R}_{\mathbf{x}'} = \mathbf{\Gamma}_{\mathbf{x}\mathbf{x}_s}\mathbf{R}_{\mathbf{x}_s}\mathbf{\Gamma}_{\mathbf{x}\mathbf{x}_s}^T$ and $\mathbf{R}_{\mathbf{x}_i}$ are the correlation matrices of $\mathbf{x}'(k)$ and $\mathbf{x}_i(k)$, respectively, and

$$\mathbf{R}_{\text{in}} = \mathbf{R}_{\mathbf{x}_i} + \mathbf{R}_\mathbf{v} \tag{4.67}$$

is the interference-plus-noise correlation matrix. Then, our objective is to estimate $\mathbf{x}_s(k)$ or $\mathbf{x}'(k)$ from $\mathbf{y}(k)$.[1]

As before, we have

$$\mathbf{R}_{\mathbf{x}'} = \mathbf{Q}_{\mathbf{x}'}\mathbf{\Lambda}_{\mathbf{x}'}\mathbf{Q}_{\mathbf{x}'}^T, \tag{4.68}$$

where

$$\begin{aligned} \mathbf{Q}_{\mathbf{x}'} &= \left[\mathbf{q}_{\mathbf{x}',1}\ \mathbf{q}_{\mathbf{x}',2}\ \cdots\ \mathbf{q}_{\mathbf{x}',L}\right] \\ &= \left[\mathbf{T}_{\mathbf{x}'}\ \mathbf{\Xi}_{\mathbf{x}'}\right], \end{aligned} \tag{4.69}$$

is an orthogonal matrix, with $\mathbf{T}_{\mathbf{x}'}$ and $\mathbf{\Xi}_{\mathbf{x}'}$ being matrices of sizes $L \times P$ and $L \times (L-P)$, respectively, and

$$\mathbf{\Lambda}_{\mathbf{x}'} = \text{diag}\left(\lambda_{\mathbf{x}',1}, \lambda_{\mathbf{x}',2}, \ldots, \lambda_{\mathbf{x}',L}\right) \tag{4.70}$$

---

[1] To simplify the presentation, we consider $\mathbf{x}'(k)$ as the desired signal vector. However, from the estimator, only the first P components are relevant and should be considered since $\mathbf{x}_s(k)$ (of length P) is the true desired signal vector.

is a diagonal matrix, with $\lambda_{\mathbf{x}',1} \geq \lambda_{\mathbf{x}',2} \geq \cdots \geq \lambda_{\mathbf{x}',P} > 0$ and $\lambda_{\mathbf{x}',P+1} = \lambda_{\mathbf{x}',P+2} = \cdots = \lambda_{\mathbf{x}',L} = 0$. Therefore, we can express (4.65) as

$$\mathbf{y}(k) = \mathbf{T}_{\mathbf{x}'}\widetilde{\mathbf{x}}_{\mathrm{s}}(k) + \mathbf{x}_{\mathrm{i}}(k) + \mathbf{v}(k), \tag{4.71}$$

where

$$\begin{aligned} \widetilde{\mathbf{x}}_{\mathrm{s}}(k) &= \mathbf{T}_{\mathbf{x}'}^T \mathbf{x}'(k) \\ &= \mathbf{T}_{\mathbf{x}'}^T \boldsymbol{\Gamma}_{\mathbf{x}\mathbf{x}_{\mathrm{s}}} \mathbf{x}_{\mathrm{s}}(k) \end{aligned} \tag{4.72}$$

is the transformed desired signal vector of length P and the correlation matrix of $\mathbf{y}(k)$ is

$$\begin{aligned} \mathbf{R}_{\mathbf{y}} &= \mathbf{T}_{\mathbf{x}'}\mathbf{R}_{\widetilde{\mathbf{x}}_{\mathrm{s}}}\mathbf{T}_{\mathbf{x}'}^T + \mathbf{R}_{\mathrm{in}} \\ &= \mathbf{T}_{\mathbf{x}'}\boldsymbol{\Lambda}_{\widetilde{\mathbf{x}}_{\mathrm{s}}}\mathbf{T}_{\mathbf{x}'}^T + \mathbf{R}_{\mathrm{in}}, \end{aligned} \tag{4.73}$$

where

$$\begin{aligned} \mathbf{R}_{\widetilde{\mathbf{x}}_{\mathrm{s}}} &= E\left[\widetilde{\mathbf{x}}_{\mathrm{s}}(k)\widetilde{\mathbf{x}}_{\mathrm{s}}^T(k)\right] \\ &= \mathrm{diag}\left(\lambda_{\mathbf{x}',1}, \lambda_{\mathbf{x}',2}, \ldots, \lambda_{\mathbf{x}',P}\right) \\ &= \boldsymbol{\Lambda}_{\widetilde{\mathbf{x}}_{\mathrm{s}}}. \end{aligned} \tag{4.74}$$

4.5.2 Linear Filtering with a Rectangular Matrix

The transformed desired signal vector, $\widetilde{\mathbf{x}}_{\mathrm{s}}(k)$, is estimated as before, i.e.,

$$\begin{aligned} \widetilde{\mathbf{z}}_{\mathrm{s}}(k) &= \widetilde{\mathbf{H}}\mathbf{y}(k) \\ &= \widetilde{\mathbf{x}}_{\mathrm{fd}}(k) + \widetilde{\mathbf{x}}_{\mathrm{ri}}(k) + \widetilde{\mathbf{v}}_{\mathrm{rn}}(k), \end{aligned} \tag{4.75}$$

where $\widetilde{\mathbf{H}}$ is a rectangular filtering matrix of size $P \times L$,

$$\widetilde{\mathbf{x}}_{\mathrm{fd}}(k) = \widetilde{\mathbf{H}}\mathbf{T}_{\mathbf{x}'}\widetilde{\mathbf{x}}_{\mathrm{s}}(k) \tag{4.76}$$

is the filtered transformed desired signal,

$$\widetilde{\mathbf{x}}_{\mathrm{ri}}(k) = \widetilde{\mathbf{H}}\mathbf{x}_{\mathrm{i}}(k) \tag{4.77}$$

is the residual interference, and

$$\widetilde{\mathbf{v}}_{\mathrm{rn}}(k) = \widetilde{\mathbf{H}}\mathbf{v}(k) \tag{4.78}$$

is the residual noise. As a result, the estimate of $\mathbf{x}'(k)$ is supposed to be[2]

$$\begin{aligned}\mathbf{z}'(k) &= \mathbf{T}_{\mathbf{x}'}\widetilde{\mathbf{z}}_{\mathrm{s}}(k) \\ &= \mathbf{T}_{\mathbf{x}'}\widetilde{\mathbf{H}}\mathbf{y}(k) \\ &= \mathbf{H}\mathbf{y}(k),\end{aligned} \tag{4.79}$$

where

$$\mathbf{H} = \mathbf{T}_{\mathbf{x}'}\widetilde{\mathbf{H}}. \tag{4.80}$$

The correlation matrix of $\widetilde{\mathbf{z}}_{\mathrm{s}}(k)$ is then

$$\mathbf{R}_{\widetilde{\mathbf{z}}_{\mathrm{s}}} = \mathbf{R}_{\widetilde{\mathbf{x}}_{\mathrm{fd}}} + \mathbf{R}_{\widetilde{\mathbf{x}}_{\mathrm{ri}}} + \mathbf{R}_{\widetilde{\mathbf{v}}_{\mathrm{rn}}}, \tag{4.81}$$

where

$$\mathbf{R}_{\widetilde{\mathbf{x}}_{\mathrm{fd}}} = \widetilde{\mathbf{H}}\mathbf{T}_{\mathbf{x}'}\mathbf{\Lambda}_{\widetilde{\mathbf{x}}_{\mathrm{s}}}\mathbf{T}_{\mathbf{x}'}^T\widetilde{\mathbf{H}}^T, \tag{4.82}$$

$$\mathbf{R}_{\widetilde{\mathbf{x}}_{\mathrm{ri}}} = \widetilde{\mathbf{H}}\mathbf{R}_{\mathbf{x}_{\mathrm{i}}}\widetilde{\mathbf{H}}^T, \tag{4.83}$$

$$\mathbf{R}_{\widetilde{\mathbf{v}}_{\mathrm{rn}}} = \widetilde{\mathbf{H}}\mathbf{R}_{\mathbf{v}}\widetilde{\mathbf{H}}^T. \tag{4.84}$$

4.5.3 Performance Measures

The definition of the input SNR is slightly different from (4.24). Indeed, it should be computed from (4.65), since we consider $\mathbf{x}_{\mathrm{i}}(k)$ as an interference and $\mathbf{x}'(k)$ as the desired signal vector. We obtain

$$\mathrm{iSNR} = \frac{\mathrm{tr}\left(\mathbf{R}_{\mathbf{x}'}\right)}{\mathrm{tr}\left(\mathbf{R}_{\mathrm{in}}\right)}. \tag{4.85}$$

The output SNR and the noise reduction factor are, respectively,

$$\begin{aligned}\mathrm{oSNR}\left(\widetilde{\mathbf{H}}\right) &= \frac{\mathrm{tr}\left(\mathbf{R}_{\widetilde{\mathbf{x}}_{\mathrm{fd}}}\right)}{\mathrm{tr}\left(\mathbf{R}_{\widetilde{\mathbf{x}}_{\mathrm{ri}}} + \mathbf{R}_{\widetilde{\mathbf{v}}_{\mathrm{rn}}}\right)} \\ &= \frac{\mathrm{tr}\left(\widetilde{\mathbf{H}}\mathbf{T}_{\mathbf{x}'}\mathbf{\Lambda}_{\widetilde{\mathbf{x}}_{\mathrm{s}}}\mathbf{T}_{\mathbf{x}'}^T\widetilde{\mathbf{H}}^T\right)}{\mathrm{tr}\left(\widetilde{\mathbf{H}}\mathbf{R}_{\mathrm{in}}\widetilde{\mathbf{H}}^T\right)}\end{aligned} \tag{4.86}$$

and

$$\xi_{\mathrm{nr}}\left(\widetilde{\mathbf{H}}\right) = \frac{\mathrm{tr}\left(\mathbf{R}_{\mathrm{in}}\right)}{\mathrm{tr}\left(\widetilde{\mathbf{H}}\mathbf{R}_{\mathrm{in}}\widetilde{\mathbf{H}}^T\right)}. \tag{4.87}$$

---

[2]Eventually, only the first P components of $\mathbf{z}'(k)$ are relevant.

To quantify distortion, we define the speech reduction factor and the speech distortion index as, respectively,

$$\begin{aligned}\xi_{\text{sr}}\left(\widetilde{\mathbf{H}}\right) &= \frac{\text{tr}\left(\mathbf{R}_{\mathbf{x}'}\right)}{\text{tr}\left(\mathbf{R}_{\widetilde{\mathbf{x}}_{\text{fd}}}\right)} \\ &= \frac{\text{tr}\left(\boldsymbol{\Lambda}_{\widetilde{\mathbf{x}}_{\text{s}}}\right)}{\text{tr}\left(\widetilde{\mathbf{H}}\mathbf{T}_{\mathbf{x}'}\boldsymbol{\Lambda}_{\widetilde{\mathbf{x}}_{\text{s}}}\mathbf{T}_{\mathbf{x}'}^T\widetilde{\mathbf{H}}^T\right)}\end{aligned} \tag{4.88}$$

and

$$\begin{aligned}\upsilon_{\text{sd}}\left(\widetilde{\mathbf{H}}\right) &= \frac{E\left\{\left[\widetilde{\mathbf{x}}_{\text{fd}}(k) - \widetilde{\mathbf{x}}_{\text{s}}(k)\right]^T\left[\widetilde{\mathbf{x}}_{\text{fd}}(k) - \widetilde{\mathbf{x}}_{\text{s}}(k)\right]\right\}}{\text{tr}\left(\mathbf{R}_{\widetilde{\mathbf{x}}_{\text{s}}}\right)} \\ &= \frac{\text{tr}\left[\left(\widetilde{\mathbf{H}}\mathbf{T}_{\mathbf{x}'} - \mathbf{I}_P\right)\boldsymbol{\Lambda}_{\widetilde{\mathbf{x}}_{\text{s}}}\left(\widetilde{\mathbf{H}}\mathbf{T}_{\mathbf{x}'} - \mathbf{I}_P\right)^T\right]}{\text{tr}\left(\boldsymbol{\Lambda}_{\widetilde{\mathbf{x}}_{\text{s}}}\right)}.\end{aligned} \tag{4.89}$$

Of course, the relation given in (4.29) is also verified in this context.

The error signal vector between the estimated and desired signals is

$$\begin{aligned}\widetilde{\mathbf{e}}(k) &= \widetilde{\mathbf{z}}_{\text{s}}(k) - \widetilde{\mathbf{x}}_{\text{s}}(k) \\ &= \widetilde{\mathbf{e}}_{\text{ds}}(k) + \widetilde{\mathbf{e}}_{\text{rs}}(k),\end{aligned} \tag{4.90}$$

where

$$\begin{aligned}\widetilde{\mathbf{e}}_{\text{ds}}(k) &= \widetilde{\mathbf{x}}_{\text{fd}}(k) - \widetilde{\mathbf{x}}_{\text{s}}(k) \\ &= \left(\widetilde{\mathbf{H}}\mathbf{T}_{\mathbf{x}'} - \mathbf{I}_P\right)\widetilde{\mathbf{x}}_{\text{s}}(k)\end{aligned} \tag{4.91}$$

is the signal distortion due to the rectangular filtering matrix and

$$\begin{aligned}\widetilde{\mathbf{e}}_{\text{rs}}(k) &= \widetilde{\mathbf{x}}_{\text{ri}}(k) + \widetilde{\mathbf{v}}_{\text{rn}}(k) \\ &= \widetilde{\mathbf{H}}\mathbf{x}_{\text{i}}(k) + \widetilde{\mathbf{H}}\mathbf{v}(k)\end{aligned} \tag{4.92}$$

represents the residual interference-plus-noise. The MSE criterion is then

$$\begin{aligned}J\left(\widetilde{\mathbf{H}}\right) &= \text{tr}\left\{E\left[\widetilde{\mathbf{e}}(k)\widetilde{\mathbf{e}}^T(k)\right]\right\} \\ &= \text{tr}\left(\boldsymbol{\Lambda}_{\widetilde{\mathbf{x}}_{\text{s}}}\right) + \text{tr}\left(\widetilde{\mathbf{H}}\mathbf{R}_{\mathbf{y}}\widetilde{\mathbf{H}}^T\right) - 2\text{tr}\left(\widetilde{\mathbf{H}}\mathbf{T}_{\mathbf{x}'}\boldsymbol{\Lambda}_{\widetilde{\mathbf{x}}_{\text{s}}}\right) \\ &= J_{\text{ds}}\left(\widetilde{\mathbf{H}}\right) + J_{\text{rs}}\left(\widetilde{\mathbf{H}}\right),\end{aligned} \tag{4.93}$$

where

$$J_{\text{ds}}\left(\widetilde{\mathbf{H}}\right) = \text{tr}\left\{E\left[\widetilde{\mathbf{e}}_{\text{ds}}(k)\widetilde{\mathbf{e}}_{\text{ds}}^T(k)\right]\right\}$$

$$= \text{tr}\left[\left(\widetilde{\mathbf{H}}\mathbf{T}_{\mathbf{x}'} - \mathbf{I}_P\right) \boldsymbol{\Lambda}_{\widetilde{\mathbf{x}}_s} \left(\widetilde{\mathbf{H}}\mathbf{T}_{\mathbf{x}'} - \mathbf{I}_P\right)^T\right]$$
$$= \text{tr}\left(\boldsymbol{\Lambda}_{\widetilde{\mathbf{x}}_s}\right) \upsilon_{\text{sd}}\left(\widetilde{\mathbf{H}}\right) \tag{4.94}$$

and

$$\begin{aligned} J_{\text{rs}}\left(\widetilde{\mathbf{H}}\right) &= \text{tr}\left\{E\left[\widetilde{\mathbf{e}}_{\text{rs}}(k)\widetilde{\mathbf{e}}_{\text{rs}}^T(k)\right]\right\} \\ &= \text{tr}\left(\widetilde{\mathbf{H}}\mathbf{R}_{\text{in}}\widetilde{\mathbf{H}}^T\right) \\ &= \frac{\text{tr}\left(\mathbf{R}_{\text{in}}\right)}{\xi_{\text{nr}}\left(\widetilde{\mathbf{H}}\right)}. \end{aligned} \tag{4.95}$$

4.5.4 Optimal Rectangular Filtering Matrices

It can be shown that the maximum SNR filtering matrix is given by

$$\widetilde{\mathbf{H}}_{\max} = \begin{bmatrix} \varsigma_1 \mathbf{b}_{\max}^T \\ \varsigma_2 \mathbf{b}_{\max}^T \\ \vdots \\ \varsigma_P \mathbf{b}_{\max}^T \end{bmatrix}, \tag{4.96}$$

where $\mathbf{b}_{\max}$ is the eigenvector corresponding to the maximum eigenvalue, $\lambda_{\max}\left(\mathbf{R}_{\text{in}}^{-1}\mathbf{R}_{\mathbf{x}'}\right)$, of the matrix $\mathbf{R}_{\text{in}}^{-1}\mathbf{R}_{\mathbf{x}'}$ and

$$\begin{aligned} \varsigma_p &= \frac{\lambda_{\mathbf{x}',p}\mathbf{b}_{\max}^T\mathbf{q}_{\mathbf{x}',p}}{\mathbf{b}_{\max}^T\mathbf{R}_{\mathbf{x}'}\mathbf{b}_{\max}} \\ &= \frac{\lambda_{\mathbf{x}',p}\mathbf{b}_{\max}^T\mathbf{q}_{\mathbf{x}',p}}{\lambda_{\max}\left(\mathbf{R}_{\text{in}}^{-1}\mathbf{R}_{\mathbf{x}'}\right)}, \quad p = 1, 2, \ldots, P. \end{aligned} \tag{4.97}$$

We can rewrite (4.96) as

$$\widetilde{\mathbf{H}}_{\max} = \boldsymbol{\Lambda}_{\widetilde{\mathbf{x}}_s}\mathbf{T}_{\mathbf{x}'}^T \frac{\mathbf{b}_{\max}\mathbf{b}_{\max}^T}{\lambda_{\max}\left(\mathbf{R}_{\text{in}}^{-1}\mathbf{R}_{\mathbf{x}'}\right)}. \tag{4.98}$$

We deduce that the maximum SNR filtering matrix for the estimation of $\mathbf{x}'(k)$ is

$$\mathbf{H}_{\max} = \mathbf{R}_{\mathbf{x}'} \frac{\mathbf{b}_{\max}\mathbf{b}_{\max}^T}{\lambda_{\max}\left(\mathbf{R}_{\text{in}}^{-1}\mathbf{R}_{\mathbf{x}'}\right)}. \tag{4.99}$$

It is easy to derive the Wiener filtering matrix:

$$\begin{aligned}\widetilde{\mathbf{H}}_{\mathrm{W}} &= \mathbf{\Lambda}_{\widetilde{\mathbf{x}}_{\mathrm{s}}}\mathbf{T}_{\mathbf{x}'}^T\mathbf{R}_{\mathbf{y}}^{-1} \\ &= \left(\mathbf{\Lambda}_{\widetilde{\mathbf{x}}_{\mathrm{s}}}^{-1} + \mathbf{T}_{\mathbf{x}'}^T\mathbf{R}_{\mathrm{in}}^{-1}\mathbf{T}_{\mathbf{x}'}\right)^{-1}\mathbf{T}_{\mathbf{x}'}^T\mathbf{R}_{\mathrm{in}}^{-1}. \end{aligned} \tag{4.100}$$

We deduce that the equivalent Wiener filtering matrix for the estimation of the vector $\mathbf{x}'(k)$ is

$$\begin{aligned}\mathbf{H}_{\mathrm{W}} &= \mathbf{T}_{\mathbf{x}'}\widetilde{\mathbf{H}}_{\mathrm{W}} \\ &= \mathbf{R}_{\mathbf{x}'}\mathbf{R}_{\mathbf{y}}^{-1} \\ &= \mathbf{I}_L - \mathbf{R}_{\mathrm{in}}\mathbf{R}_{\mathbf{y}}^{-1}. \end{aligned} \tag{4.101}$$

This reduced-rank Wiener filtering matrix is different from the classical Wiener filtering matrix [1].

The MVDR filtering matrix is

$$\begin{aligned}\widetilde{\mathbf{H}}_{\mathrm{MVDR}} &= \left(\mathbf{T}_{\mathbf{x}'}^T\mathbf{R}_{\mathrm{in}}^{-1}\mathbf{T}_{\mathbf{x}'}\right)^{-1}\mathbf{T}_{\mathbf{x}'}^T\mathbf{R}_{\mathrm{in}}^{-1} \\ &= \left(\mathbf{T}_{\mathbf{x}'}^T\mathbf{R}_{\mathbf{y}}^{-1}\mathbf{T}_{\mathbf{x}'}\right)^{-1}\mathbf{T}_{\mathbf{x}'}^T\mathbf{R}_{\mathbf{y}}^{-1} \end{aligned} \tag{4.102}$$

for the estimation of $\widetilde{\mathbf{x}}_{\mathrm{s}}(k)$ and

$$\begin{aligned}\mathbf{H}_{\mathrm{MVDR}} &= \mathbf{T}_{\mathbf{x}'}\left(\mathbf{T}_{\mathbf{x}'}^T\mathbf{R}_{\mathrm{in}}^{-1}\mathbf{T}_{\mathbf{x}'}\right)^{-1}\mathbf{T}_{\mathbf{x}'}^T\mathbf{R}_{\mathrm{in}}^{-1} \\ &= \mathbf{T}_{\mathbf{x}'}\left(\mathbf{T}_{\mathbf{x}'}^T\mathbf{R}_{\mathbf{y}}^{-1}\mathbf{T}_{\mathbf{x}'}\right)^{-1}\mathbf{T}_{\mathbf{x}'}^T\mathbf{R}_{\mathbf{y}}^{-1} \end{aligned} \tag{4.103}$$

for the estimation of $\mathbf{x}'(k)$.

It is not hard to derive the tradeoff filtering matrix:

$$\widetilde{\mathbf{H}}_{\mathrm{T},\mu} = \mathbf{\Lambda}_{\widetilde{\mathbf{x}}_{\mathrm{s}}}\mathbf{T}_{\mathbf{x}'}^T\left(\mathbf{R}_{\mathbf{x}'} + \mu\mathbf{R}_{\mathrm{in}}\right)^{-1}, \tag{4.104}$$

which can be rewritten, thanks to the Woodbury's identity, as

$$\widetilde{\mathbf{H}}_{\mathrm{T},\mu} = \left(\mu\mathbf{\Lambda}_{\widetilde{\mathbf{x}}_{\mathrm{s}}}^{-1} + \mathbf{T}_{\mathbf{x}'}^T\mathbf{R}_{\mathrm{in}}^{-1}\mathbf{T}_{\mathbf{x}'}\right)^{-1}\mathbf{T}_{\mathbf{x}'}^T\mathbf{R}_{\mathrm{in}}^{-1}, \tag{4.105}$$

where $\mu > 0$ in (4.104) and $\mu \geq 0$ in (4.105). Therefore,

$$\begin{aligned}\mathbf{H}_{\mathrm{T},\mu} &= \mathbf{T}_{\mathbf{x}'}\left(\mu\mathbf{\Lambda}_{\widetilde{\mathbf{x}}_{\mathrm{s}}}^{-1} + \mathbf{T}_{\mathbf{x}'}^T\mathbf{R}_{\mathrm{in}}^{-1}\mathbf{T}_{\mathbf{x}'}\right)^{-1}\mathbf{T}_{\mathbf{x}'}^T\mathbf{R}_{\mathrm{in}}^{-1} \\ &= \mathbf{R}_{\mathbf{x}'}\left(\mathbf{R}_{\mathbf{x}'} + \mu\mathbf{R}_{\mathrm{in}}\right)^{-1}. \end{aligned} \tag{4.106}$$

As for the LCMV filtering matrix, it is given by

$$\widetilde{\mathbf{H}}_{\text{LCMV}} = \left[\mathbf{I}_P \; \mathbf{0}_{P\times Q} \right] \left(\mathbf{C}_{\mathbf{x}'\mathbf{v}}^T \mathbf{R}_{\mathbf{y}}^{-1} \mathbf{C}_{\mathbf{x}'\mathbf{v}} \right)^{-1} \mathbf{C}_{\mathbf{x}'\mathbf{v}}^T \mathbf{R}_{\mathbf{y}}^{-1} \tag{4.107}$$

or

$$\mathbf{H}_{\text{LCMV}} = \mathbf{T}_{\mathbf{x}'} \left[\mathbf{I}_P \; \mathbf{0}_{P\times Q} \right] \left(\mathbf{C}_{\mathbf{x}'\mathbf{v}}^T \mathbf{R}_{\mathbf{y}}^{-1} \mathbf{C}_{\mathbf{x}'\mathbf{v}} \right)^{-1} \mathbf{C}_{\mathbf{x}'\mathbf{v}}^T \mathbf{R}_{\mathbf{y}}^{-1}, \tag{4.108}$$

where

$$\mathbf{C}_{\mathbf{x}'\mathbf{v}} = \left[\mathbf{T}_{\mathbf{x}'} \; \mathbf{T}_{\mathbf{v}} \right]. \tag{4.109}$$

REFERENCES

[1] J. Benesty, J. Chen, Y. Huang, I. Cohen, Noise Reduction in Speech Processing, Springer-Verlag, Berlin, Germany, 2009.

[2] P. Loizou, Speech Enhancement: Theory and Practice, CRC Press, Boca Raton, FL, 2007.

[3] J. Benesty, J. Chen, Optimal time-domain noise reduction filters–a theoretical study, Springer Briefs in Electrical and Computer Engineering, Springer-Verlag, 2011.

[4] Y. Ephraim, H.L. Van Trees, A signal subspace approach for speech enhancement, IEEE Trans. Speech Audio Process. 3 (1995) 251–266.

[5] P.S.K. Hansen, Signal Subspace Methods for Speech Enhancement, Ph.D. Dissertation, Technical University of Denmark, Lyngby, Denmark, 1997.

[6] S.H. Jensen, P.C. Hansen, S.D. Hansen, J.A. Sørensen, Reduction of broad-band noise in speech by truncated QSVD, IEEE Trans. Speech Audio Process. 3 (1995) 439–448.

[7] M. Dendrinos, S. Bakamidis, G. Carayannis, Speech enhancement from noise: a regenerative approach, Speech Commun. 10 (1991) 45–57.

CHAPTER 5

Multichannel Speech Enhancement in the Time Domain

In this chapter, we show how to extend the ideas of single-channel noise reduction in the time domain to the multichannel (or multisensors) case. The main advantage here is that we take the spatial information into account. This generalization is not straightforward since several steps are required to get results that look similar in their formulation to the single microphone scenario. We show again the importance of the speech subspace in all derived optimal filtering matrix for multichannel speech enhancement.

5.1 SIGNAL MODEL AND PROBLEM FORMULATION

We consider the conventional signal model in which a microphone array with M sensors captures a convolved source signal in some noise field. The received signals are expressed as [1,2]

$$\begin{aligned} y_m(k) &= g_m(k) * s(k) + v_m(k) \\ &= x_m(k) + v_m(k), \quad m = 1, 2, \ldots, M, \end{aligned} \tag{5.1}$$

where $g_m(k)$ is the acoustic impulse response from the unknown speech source, $s(k)$, location to the mth microphone, $*$ stands for linear convolution, and $v_m(k)$ is the additive noise at microphone m. We assume that the signals $x_m(k) = g_m(k) * s(k)$ and $v_m(k)$ are uncorrelated, zero mean, stationary, real, and broadband. By definition, the convolved speech signals, $x_m(k), m = 1, 2, \ldots, M$, are coherent while the noise signals, $v_m(k), m = 1, 2, \ldots, M$, are typically only partially coherent. We assume that the microphone signals are aligned so that the array looks in the direction of the source, which is considered to be known or can be estimated. This preprocessing step may be needed in order to avoid using very large filtering matrices.

By processing the data by blocks of L samples, the signal model given in (5.1) can be put into a vector form as

$$\mathbf{y}_m(k) = \mathbf{x}_m(k) + \mathbf{v}_m(k), \quad m = 1, 2, \ldots, M, \tag{5.2}$$

Speech Enhancement. http://dx.doi.org/10.1016/B978-0-12-800139-4.00005-0

where

$$\mathbf{y}_m(k) = \left[y_m(k)\ y_m(k-1) \cdots y_m(k-L+1) \right]^T \quad (5.3)$$

is a vector of length L, and $\mathbf{x}_m(k)$ and $\mathbf{v}_m(k)$ are defined similarly to $\mathbf{y}_m(k)$ from (5.3). It is more convenient to concatenate the M vectors $\mathbf{y}_m(k)$ together as

$$\begin{aligned} \underline{\mathbf{y}}(k) &= \left[\mathbf{y}_1^T(k)\ \mathbf{y}_2^T(k) \cdots \mathbf{y}_M^T(k) \right]^T \\ &= \underline{\mathbf{x}}(k) + \underline{\mathbf{v}}(k), \end{aligned} \quad (5.4)$$

where the vectors $\underline{\mathbf{x}}(k)$ and $\underline{\mathbf{v}}(k)$, of length ML, are defined in a similar way to $\underline{\mathbf{y}}(k)$. Since $x_m(k)$ and $v_m(k)$ are uncorrelated by assumption, the correlation matrix (of size $ML \times ML$) of the microphone signals is

$$\begin{aligned} \mathbf{R}_{\underline{\mathbf{y}}} &= E\left[\underline{\mathbf{y}}(k)\underline{\mathbf{y}}^T(k) \right] \\ &= \mathbf{R}_{\underline{\mathbf{x}}} + \mathbf{R}_{\underline{\mathbf{v}}}, \end{aligned} \quad (5.5)$$

where $\mathbf{R}_{\underline{\mathbf{x}}} = E\left[\underline{\mathbf{x}}(k)\underline{\mathbf{x}}^T(k) \right]$ and $\mathbf{R}_{\underline{\mathbf{v}}} = E\left[\underline{\mathbf{v}}(k)\underline{\mathbf{v}}^T(k) \right]$ are the correlation matrices of $\underline{\mathbf{x}}(k)$ and $\underline{\mathbf{v}}(k)$, respectively.

In this work, our desired signal vector is designated by the clean (but convolved) speech signal samples received at microphone 1, namely $\mathbf{x}_1(k)$. Obviously, any vector $\mathbf{x}_m(k)$ could be taken as the reference. Our problem then may be stated as follows: given M mixtures of two uncorrelated signal vectors $\mathbf{x}_m(k)$ and $\mathbf{v}_m(k)$, our aim is to preserve $\mathbf{x}_1(k)$ while minimizing the contribution of the noise terms, $\mathbf{v}_m(k)$, $m = 1, 2, \ldots, M$, at the array output.

Since $\mathbf{x}_1(k)$ is the desired signal vector, we need to write it as a function of $\underline{\mathbf{x}}(k)$. Indeed, by decomposing $\underline{\mathbf{x}}(k)$ into two orthogonal components, one proportional to the desired signal vector, $\mathbf{x}_1(k)$, and the other corresponding to the interference, we get [3]

$$\begin{aligned} \underline{\mathbf{x}}(k) &= \mathbf{R}_{\underline{\mathbf{x}}\mathbf{x}_1}\mathbf{R}_{\mathbf{x}_1}^{-1}\mathbf{x}_1(k) + \underline{\mathbf{x}}_{\mathrm{i}}(k) \\ &= \mathbf{\Gamma}_{\underline{\mathbf{x}}\mathbf{x}_1}\mathbf{x}_1(k) + \underline{\mathbf{x}}_{\mathrm{i}}(k) \\ &= \underline{\mathbf{x}}'(k) + \underline{\mathbf{x}}_{\mathrm{i}}(k), \end{aligned} \quad (5.6)$$

where

$$\mathbf{\Gamma}_{\underline{\mathbf{x}}\mathbf{x}_1} = \mathbf{R}_{\underline{\mathbf{x}}\mathbf{x}_1}\mathbf{R}_{\mathbf{x}_1}^{-1} \quad (5.7)$$

is the time-domain steering matrix, $\mathbf{R}_{\underline{\mathbf{x}}\mathbf{x}_1} = E\left[\underline{\mathbf{x}}(k)\mathbf{x}_1^T(k) \right]$ is the cross-correlation matrix of size $ML \times L$ between $\underline{\mathbf{x}}(k)$ and $\mathbf{x}_1(k)$,

$\mathbf{R}_{\mathbf{x}_1} = E\left[\mathbf{x}_1(k)\mathbf{x}_1^T(k)\right]$ is the correlation matrix (of size $L \times L$) of $\mathbf{x}_1(k)$, $\underline{\mathbf{x}}'(k) = \mathbf{\Gamma}_{\underline{\mathbf{x}}\mathbf{x}_1}\mathbf{x}_1(k)$, and $\underline{\mathbf{x}}_{\mathrm{i}}(k)$ is the interference signal vector. Obviously, the first L components of $\underline{\mathbf{x}}'(k)$ are equal to $\mathbf{x}_1(k)$, while the first L components of $\underline{\mathbf{x}}_{\mathrm{i}}(k)$ are equal to $\mathbf{0}_{L\times 1}$. Therefore, we can estimate $\underline{\mathbf{x}}'(k)$ instead of $\mathbf{x}_1(k)$ and then take the first L components of the corresponding estimator; hence, $\underline{\mathbf{x}}'(k)$ is also considered as the desired signal vector. It can be verified that $\underline{\mathbf{x}}'(k)$ and $\underline{\mathbf{x}}_{\mathrm{i}}(k)$ are orthogonal, i.e.,

$$E\left[\underline{\mathbf{x}}'(k)\underline{\mathbf{x}}_{\mathrm{i}}^T(k)\right] = \mathbf{0}_{ML\times ML}. \tag{5.8}$$

Therefore, (5.4) can be rewritten as

$$\underline{\mathbf{y}}(k) = \underline{\mathbf{x}}'(k) + \underline{\mathbf{x}}_{\mathrm{i}}(k) + \underline{\mathbf{v}}(k) \tag{5.9}$$

and the correlation matrix of $\underline{\mathbf{y}}(k)$ from the previous expression is

$$\begin{aligned} \mathbf{R}_{\underline{\mathbf{y}}} &= \mathbf{R}_{\underline{\mathbf{x}}'} + \mathbf{R}_{\underline{\mathbf{x}}_{\mathrm{i}}} + \mathbf{R}_{\underline{\mathbf{v}}} \\ &= \mathbf{R}_{\underline{\mathbf{x}}'} + \mathbf{R}_{\mathrm{in}}, \end{aligned} \tag{5.10}$$

where $\mathbf{R}_{\underline{\mathbf{x}}'} = \mathbf{\Gamma}_{\underline{\mathbf{x}}\mathbf{x}_1}\mathbf{R}_{\mathbf{x}_1}\mathbf{\Gamma}_{\underline{\mathbf{x}}\mathbf{x}_1}^T$ and $\mathbf{R}_{\underline{\mathbf{x}}_{\mathrm{i}}}$ are the correlation matrices of $\underline{\mathbf{x}}'(k)$ and $\underline{\mathbf{x}}_{\mathrm{i}}(k)$, respectively, and

$$\mathbf{R}_{\mathrm{in}} = \mathbf{R}_{\underline{\mathbf{x}}_{\mathrm{i}}} + \mathbf{R}_{\underline{\mathbf{v}}} \tag{5.11}$$

is the interference-plus-noise correlation matrix.

We can decompose $\mathbf{R}_{\underline{\mathbf{x}}'}$ as follows:

$$\mathbf{R}_{\underline{\mathbf{x}}'} = \mathbf{Q}_{\underline{\mathbf{x}}'}\mathbf{\Lambda}_{\underline{\mathbf{x}}'}\mathbf{Q}_{\underline{\mathbf{x}}'}^T, \tag{5.12}$$

where

$$\mathbf{Q}_{\underline{\mathbf{x}}'} = \left[\mathbf{q}_{\underline{\mathbf{x}}',1}\ \mathbf{q}_{\underline{\mathbf{x}}',2}\ \cdots\ \mathbf{q}_{\underline{\mathbf{x}}',ML}\right] \tag{5.13}$$

is an orthogonal matrix, i.e., $\mathbf{Q}_{\underline{\mathbf{x}}'}^T\mathbf{Q}_{\underline{\mathbf{x}}'} = \mathbf{Q}_{\underline{\mathbf{x}}'}\mathbf{Q}_{\underline{\mathbf{x}}'}^T = \mathbf{I}_{ML}$, with $\mathbf{I}_{ML}$ being the $ML \times ML$ identity matrix, and

$$\mathbf{\Lambda}_{\underline{\mathbf{x}}'} = \mathrm{diag}\left(\lambda_{\underline{\mathbf{x}}',1}, \lambda_{\underline{\mathbf{x}}',2}, \ldots, \lambda_{\underline{\mathbf{x}}',ML}\right) \tag{5.14}$$

is a diagonal matrix, with $\lambda_{\underline{\mathbf{x}}',1} \geq \lambda_{\underline{\mathbf{x}}',2} \geq \cdots \geq \lambda_{\underline{\mathbf{x}}',L} > 0$ and $\lambda_{\underline{\mathbf{x}}',L+1} = \lambda_{\underline{\mathbf{x}}',L+2} = \cdots = \lambda_{\underline{\mathbf{x}}',ML} = 0$. It is of interest to decompose the orthogonal matrix as

$$\mathbf{Q}_{\underline{\mathbf{x}}'} = \left[\mathbf{T}_{\underline{\mathbf{x}}'}\ \mathbf{\Xi}_{\underline{\mathbf{x}}'}\right], \tag{5.15}$$

where the $ML \times L$ matrix $\mathbf{T}_{\underline{\mathbf{x}}'}$ contains the eigenvectors corresponding to the nonzero eigenvalues of $\mathbf{R}_{\underline{\mathbf{x}}'}$ and the $ML \times (ML - L)$ matrix $\mathbf{\Xi}_{\underline{\mathbf{x}}'}$ contains the eigenvectors corresponding to the null eigenvalues of $\mathbf{R}_{\underline{\mathbf{x}}'}$. Therefore,

$$\begin{aligned}\underline{\mathbf{x}}'(k) &= \mathbf{Q}_{\underline{\mathbf{x}}'}\mathbf{Q}_{\underline{\mathbf{x}}'}^T\underline{\mathbf{x}}'(k)\\ &= \mathbf{T}_{\underline{\mathbf{x}}'}\mathbf{T}_{\underline{\mathbf{x}}'}^T\underline{\mathbf{x}}'(k)\\ &= \mathbf{T}_{\underline{\mathbf{x}}'}\mathbf{x}'(k), \end{aligned} \tag{5.16}$$

where

$$\mathbf{x}'(k) = \mathbf{T}_{\underline{\mathbf{x}}'}^T\underline{\mathbf{x}}'(k) \tag{5.17}$$

is the transformed desired signal vector of length L. We deduce that the signal model for noise reduction is

$$\underline{\mathbf{y}}(k) = \mathbf{T}_{\underline{\mathbf{x}}'}\mathbf{x}'(k) + \underline{\mathbf{x}}_{\mathrm{i}}(k) + \underline{\mathbf{v}}(k), \tag{5.18}$$

for which its correlation matrix is

$$\begin{aligned}\mathbf{R}_{\underline{\mathbf{y}}} &= \mathbf{T}_{\underline{\mathbf{x}}'}\mathbf{R}_{\mathbf{x}'}\mathbf{T}_{\underline{\mathbf{x}}'}^T + \mathbf{R}_{\mathrm{in}}\\ &= \mathbf{T}_{\underline{\mathbf{x}}'}\mathbf{\Lambda}_{\mathbf{x}'}\mathbf{T}_{\underline{\mathbf{x}}'}^T + \mathbf{R}_{\mathrm{in}}, \end{aligned} \tag{5.19}$$

where

$$\begin{aligned}\mathbf{R}_{\mathbf{x}'} &= E\left[\mathbf{x}'(k)\mathbf{x}'^T(k)\right]\\ &= \mathrm{diag}\left(\lambda_{\underline{\mathbf{x}}',1}, \lambda_{\underline{\mathbf{x}}',2}, \ldots, \lambda_{\underline{\mathbf{x}}',L}\right)\\ &= \mathbf{\Lambda}_{\mathbf{x}'} \end{aligned} \tag{5.20}$$

and $\mathbf{R}_{\underline{\mathbf{x}}'} = \mathbf{T}_{\underline{\mathbf{x}}'}\mathbf{\Lambda}_{\mathbf{x}'}\mathbf{T}_{\underline{\mathbf{x}}'}^T$.

5.2 LINEAR FILTERING WITH A RECTANGULAR MATRIX

Multichannel noise reduction is performed by applying a linear transformation to $\underline{\mathbf{y}}(k)$. We get

$$\begin{aligned}\mathbf{z}'(k) &= \underline{\mathbf{H}}'\underline{\mathbf{y}}(k)\\ &= \mathbf{x}'_{\mathrm{fd}}(k) + \mathbf{x}'_{\mathrm{ri}}(k) + \mathbf{v}'_{\mathrm{rn}}(k), \end{aligned} \tag{5.21}$$

where $\mathbf{z}'(k)$ is the estimate of $\mathbf{x}'(k)$,

$$\underline{\mathbf{H}}' = \begin{bmatrix} \underline{\mathbf{h}}_1'^T \\ \underline{\mathbf{h}}_2'^T \\ \vdots \\ \underline{\mathbf{h}}_L'^T \end{bmatrix} \tag{5.22}$$

is a rectangular filtering matrix of size $L \times ML$, $\underline{\mathbf{h}}_l'$, $l = 1, 2, \ldots, L$ are real-valued filters of length ML,

$$\mathbf{x}_{\mathrm{fd}}'(k) = \underline{\mathbf{H}}'\mathbf{T}_{\underline{\mathbf{x}}'}\mathbf{x}'(k) \tag{5.23}$$

is the filtered transformed desired signal,

$$\mathbf{x}_{\mathrm{ri}}'(k) = \underline{\mathbf{H}}'\underline{\mathbf{x}}_{\mathrm{i}}(k) \tag{5.24}$$

is the residual interference, and

$$\mathbf{v}_{\mathrm{rn}}'(k) = \underline{\mathbf{H}}'\underline{\mathbf{v}}(k) \tag{5.25}$$

is the residual noise. We deduce from (5.16) and (5.21) that the estimate of $\underline{\mathbf{x}}'(k)$ is

$$\begin{aligned} \underline{\mathbf{z}}'(k) &= \mathbf{T}_{\underline{\mathbf{x}}'}\mathbf{z}'(k) \\ &= \mathbf{T}_{\underline{\mathbf{x}}'}\underline{\mathbf{H}}'\underline{\mathbf{y}}(k) \end{aligned} \tag{5.26}$$

and the estimate of $\mathbf{x}_1(k)$ is

$$\begin{aligned} \mathbf{z}_1(k) &= \left[\mathbf{I}_L \ \mathbf{0}_{L\times(ML-L)} \right] \underline{\mathbf{z}}'(k) \\ &= \underline{\mathbf{H}}\underline{\mathbf{y}}(k), \end{aligned} \tag{5.27}$$

where

$$\underline{\mathbf{H}} = \underline{\mathbf{I}}_{10}\mathbf{T}_{\underline{\mathbf{x}}'}\underline{\mathbf{H}}' \tag{5.28}$$

and

$$\underline{\mathbf{I}}_{10} = \left[\mathbf{I}_L \ \mathbf{0}_{L\times(ML-L)} \right]. \tag{5.29}$$

The correlation matrix of $\mathbf{z}'(k)$ is then

$$\mathbf{R}_{\mathbf{z}'} = \mathbf{R}_{\mathbf{x}_{\mathrm{fd}}'} + \mathbf{R}_{\mathbf{x}_{\mathrm{ri}}'} + \mathbf{R}_{\mathbf{v}_{\mathrm{rn}}'}, \tag{5.30}$$

where

$$\mathbf{R}_{\mathbf{x}_{\mathrm{fd}}'} = \underline{\mathbf{H}}'\mathbf{T}_{\underline{\mathbf{x}}'}\boldsymbol{\Lambda}_{\mathbf{x}'}\mathbf{T}_{\underline{\mathbf{x}}'}^T\underline{\mathbf{H}}'^T, \tag{5.31}$$

$$\mathbf{R}_{\mathbf{x}_{\mathrm{ri}}'} = \underline{\mathbf{H}}'\mathbf{R}_{\underline{\mathbf{x}}_{\mathrm{i}}}\underline{\mathbf{H}}'^T, \tag{5.32}$$

$$\mathbf{R}_{\mathbf{v}_{\mathrm{rn}}'} = \underline{\mathbf{H}}'\mathbf{R}_{\underline{\mathbf{v}}}\underline{\mathbf{H}}'^T. \tag{5.33}$$

5.3 PERFORMANCE MEASURES

We derive the performance measures in the context of multichannel noise reduction with microphone 1 as the reference. Most of the measures are given as a function of $\underline{\mathbf{H}}$ since this filtering matrix leads directly to the estimation of the desired signal vector, $\mathbf{x}_1(k)$. However, for simplicity, the MSE is given as a function of $\underline{\mathbf{H}}'$.

5.3.1 Noise Reduction

Since microphone 1 is the reference, the input SNR is computed from the first L components of $\underline{\mathbf{y}}(k)$ as defined in (5.18). We easily find that

$$\text{iSNR} = \frac{\text{tr}\left(\mathbf{R}_{\mathbf{x}_1}\right)}{\text{tr}\left(\mathbf{R}_{\mathbf{v}_1}\right)}, \tag{5.34}$$

where $\mathbf{R}_{\mathbf{v}_1}$ is the correlation matrix of $\mathbf{v}_1(k)$.

The output SNR is obtained from (5.27). It is given by

$$\text{oSNR}\left(\underline{\mathbf{H}}\right) = \frac{\text{tr}\left(\underline{\mathbf{H}}\mathbf{T}_{\underline{\mathbf{x}}'}\mathbf{\Lambda}_{\underline{\mathbf{x}}'}\mathbf{T}_{\underline{\mathbf{x}}'}^T\underline{\mathbf{H}}^T\right)}{\text{tr}\left(\underline{\mathbf{H}}\mathbf{R}_{\text{in}}\underline{\mathbf{H}}^T\right)}. \tag{5.35}$$

Then, the main objective of speech enhancement is to find an appropriate $\underline{\mathbf{H}}$ that makes the output SNR greater than the input SNR. Consequently, the quality of the noisy signal may be enhanced.

The noise reduction factor is defined as

$$\xi_{\text{nr}}\left(\underline{\mathbf{H}}\right) = \frac{\text{tr}\left(\mathbf{R}_{\mathbf{v}_1}\right)}{\text{tr}\left(\underline{\mathbf{H}}\mathbf{R}_{\text{in}}\underline{\mathbf{H}}^T\right)}. \tag{5.36}$$

For optimal filtering matrices, we should have $\xi_{\text{nr}}\left(\underline{\mathbf{H}}\right) \geq 1$.

5.3.2 Speech Distortion

The distortion of the desired signal vector can be measured with the speech reduction factor:

$$\xi_{\text{sr}}\left(\underline{\mathbf{H}}\right) = \frac{\text{tr}\left(\mathbf{R}_{\mathbf{x}_1}\right)}{\text{tr}\left(\underline{\mathbf{H}}\mathbf{T}_{\underline{\mathbf{x}}'}\mathbf{\Lambda}_{\underline{\mathbf{x}}'}\mathbf{T}_{\underline{\mathbf{x}}'}^T\underline{\mathbf{H}}^T\right)}. \tag{5.37}$$

In the distortionless case, we have $\xi_{\text{sr}}\left(\underline{\mathbf{H}}\right) = 1$. For optimal filtering matrices, we should have $\xi_{\text{sr}}\left(\underline{\mathbf{H}}\right) \geq 1$.

We can verify the fundamental relation:

$$\frac{\text{oSNR}\left(\underline{\mathbf{H}}\right)}{\text{iSNR}} = \frac{\xi_{\text{nr}}\left(\underline{\mathbf{H}}\right)}{\xi_{\text{sr}}\left(\underline{\mathbf{H}}\right)}. \quad (5.38)$$

Another more convenient way to measure the distortion is via the speech distortion index defined as

$$\upsilon_{\text{sd}}\left(\underline{\mathbf{H}}\right) = \frac{\text{tr}\left[\left(\underline{\mathbf{H}}\mathbf{T}_{\underline{\mathbf{x}}'} - \mathbf{I}_L\right)\mathbf{\Lambda}_{\mathbf{x}'}\left(\underline{\mathbf{H}}\mathbf{T}_{\underline{\mathbf{x}}'} - \mathbf{I}_L\right)^T\right]}{\text{tr}\left(\mathbf{R}_{\mathbf{x}_1}\right)}. \quad (5.39)$$

The speech distortion index is always greater than or equal to 0; so the higher is the value of $\upsilon_{\text{sd}}\left(\underline{\mathbf{H}}\right)$, the more the desired signal is distorted.

5.3.3 MSE Criterion

We define the error signal vector (of length L) between the estimated signal vector, $\mathbf{z}'(k)$, and the transformed desired signal vector, $\mathbf{x}'(k)$, as

$$\begin{aligned}\mathbf{e}'(k) &= \mathbf{z}'(k) - \mathbf{x}'(k) \\ &= \underline{\mathbf{H}}'\underline{\mathbf{y}}(k) - \mathbf{x}'(k),\end{aligned} \quad (5.40)$$

which can also be written as the sum of two uncorrelated error signal vectors:

$$\mathbf{e}'(k) = \mathbf{e}'_{\text{ds}}(k) + \mathbf{e}'_{\text{rs}}(k), \quad (5.41)$$

where

$$\begin{aligned}\mathbf{e}'_{\text{ds}}(k) &= \mathbf{x}'_{\text{fd}}(k) - \mathbf{x}'(k) \\ &= \left(\underline{\mathbf{H}}'\mathbf{T}_{\underline{\mathbf{x}}'} - \mathbf{I}_L\right)\mathbf{x}'(k)\end{aligned} \quad (5.42)$$

is the signal distortion due to the rectangular filtering matrix and

$$\begin{aligned}\mathbf{e}'_{\text{rs}}(k) &= \mathbf{x}'_{\text{ri}}(k) + \mathbf{v}'_{\text{rn}}(k) \\ &= \underline{\mathbf{H}}'\underline{\mathbf{x}}_{\text{i}}(k) + \underline{\mathbf{H}}'\underline{\mathbf{v}}(k)\end{aligned} \quad (5.43)$$

represents the residual interference-plus-noise. The MSE criterion is then

$$\begin{aligned}J\left(\underline{\mathbf{H}}'\right) &= \text{tr}\left\{E\left[\mathbf{e}'(k)\mathbf{e}'^T(k)\right]\right\} \\ &= \text{tr}\left(\mathbf{\Lambda}_{\mathbf{x}'}\right) + \text{tr}\left(\underline{\mathbf{H}}'\mathbf{R}_{\underline{\mathbf{y}}}\underline{\mathbf{H}}'^T\right) - 2\text{tr}\left(\underline{\mathbf{H}}'\mathbf{T}_{\underline{\mathbf{x}}'}\mathbf{\Lambda}_{\mathbf{x}'}\right).\end{aligned} \quad (5.44)$$

Using the fact that $E\left[\mathbf{e}'_{\text{ds}}(k)\mathbf{e}'^{T}_{\text{rs}}(k)\right] = \mathbf{0}_{L\times L}$, $J\left(\underline{\mathbf{H}}'\right)$ can be expressed as the sum of two other MSEs, i.e.,

$$\begin{aligned} J\left(\underline{\mathbf{H}}'\right) &= \text{tr}\left\{E\left[\mathbf{e}'_{\text{ds}}(k)\mathbf{e}'^{T}_{\text{ds}}(k)\right]\right\} + \text{tr}\left\{E\left[\mathbf{e}'_{\text{rs}}(k)\mathbf{e}'^{T}_{\text{rs}}(k)\right]\right\} \\ &= J_{\text{ds}}\left(\underline{\mathbf{H}}'\right) + J_{\text{rs}}\left(\underline{\mathbf{H}}'\right), \end{aligned} \tag{5.45}$$

where

$$J_{\text{ds}}\left(\underline{\mathbf{H}}'\right) = \text{tr}\left[\left(\underline{\mathbf{H}}'\mathbf{T}_{\underline{\mathbf{x}}'} - \mathbf{I}_L\right)\mathbf{\Lambda}_{\mathbf{x}'}\left(\underline{\mathbf{H}}'\mathbf{T}_{\underline{\mathbf{x}}'} - \mathbf{I}_L\right)^T\right] \tag{5.46}$$

and

$$J_{\text{rs}}\left(\underline{\mathbf{H}}'\right) = \text{tr}\left(\underline{\mathbf{H}}'\mathbf{R}_{\text{in}}\underline{\mathbf{H}}'^{T}\right). \tag{5.47}$$

5.4 OPTIMAL RECTANGULAR FILTERING MATRICES

We are now ready to derive the most important multichannel noise reduction filtering matrices in the time domain from a signal subspace perspective.

5.4.1 Maximum SNR

We can derive two slightly different versions of the maximum SNR filtering matrix. One is obtained by maximizing (5.35), while the other is obtained by maximizing

$$\text{oSNR}\left(\underline{\mathbf{H}}'\right) = \frac{\text{tr}\left(\underline{\mathbf{H}}'\mathbf{T}_{\underline{\mathbf{x}}'}\mathbf{\Lambda}_{\mathbf{x}'}\mathbf{T}^{T}_{\underline{\mathbf{x}}'}\underline{\mathbf{H}}'^{T}\right)}{\text{tr}\left(\underline{\mathbf{H}}'\mathbf{R}_{\text{in}}\underline{\mathbf{H}}'^{T}\right)}. \tag{5.48}$$

We choose to maximize (5.48) because the corresponding filtering matrix with minimum distortion is easier to derive. We can verify that

$$\underline{\mathbf{H}}'_{\max} = \begin{bmatrix} \varsigma_1 \underline{\mathbf{b}}'^{T}_1 \\ \varsigma_2 \underline{\mathbf{b}}'^{T}_1 \\ \vdots \\ \varsigma_L \underline{\mathbf{b}}'^{T}_1 \end{bmatrix}, \tag{5.49}$$

where $\varsigma_l, l = 1, 2, \ldots, L$ are arbitrary real numbers with at least one of them different from 0 and $\underline{\mathbf{b}}'_1$ is the eigenvector corresponding to the maximum eigenvalue, λ'_1, of the matrix $\mathbf{R}^{-1}_{\text{in}}\mathbf{R}_{\underline{\mathbf{x}}'}$, maximizes (5.48). We have

$$\text{oSNR}\left(\underline{\mathbf{H}}'_{\max}\right) = \lambda'_1, \tag{5.50}$$

$$\text{oSNR}\left(\underline{\mathbf{H}}'_{\max}\right) \geq \text{iSNR}, \tag{5.51}$$

and

$$0 \leq \text{oSNR}\left(\underline{\mathbf{H}}'\right) \leq \text{oSNR}\left(\underline{\mathbf{H}}'_{\max}\right), \forall \underline{\mathbf{H}}'. \tag{5.52}$$

The choice of the values of ς_l, $l = 1, 2, \ldots, L$ is important in practice; they should be found in such a way that distortion is minimized. We can rewrite the distortion-based MSE as

$$\begin{aligned} J_{\text{ds}}\left(\underline{\mathbf{H}}'\right) &= \text{tr}\left(\mathbf{\Lambda}_{\mathbf{x}'}\right) + \text{tr}\left(\underline{\mathbf{H}}'\mathbf{R}_{\underline{\mathbf{x}}'}\underline{\mathbf{H}}'^T\right) - 2\text{tr}\left(\underline{\mathbf{H}}'\mathbf{T}_{\underline{\mathbf{x}}'}\mathbf{\Lambda}_{\mathbf{x}'}\right) \\ &= \text{tr}\left(\mathbf{\Lambda}_{\mathbf{x}'}\right) + \sum_{l=1}^{L} \underline{\mathbf{h}}_l'^T \mathbf{R}_{\underline{\mathbf{x}}'}\underline{\mathbf{h}}_l' - 2\sum_{l=1}^{L} \lambda_{\underline{\mathbf{x}}',l}\underline{\mathbf{h}}_l'^T \mathbf{q}_{\underline{\mathbf{x}}',l}. \end{aligned} \tag{5.53}$$

Substituting (5.49) into (5.53), we get

$$J_{\text{ds}}\left(\underline{\mathbf{H}}'_{\max}\right) = \text{tr}\left(\mathbf{\Lambda}_{\mathbf{x}'}\right) + \underline{\mathbf{b}}_1'^T \mathbf{R}_{\underline{\mathbf{x}}'}\underline{\mathbf{b}}_1' \sum_{l=1}^{L} \varsigma_l^2 - 2\sum_{l=1}^{L} \varsigma_l \lambda_{\underline{\mathbf{x}}',l}\underline{\mathbf{b}}_1'^T \mathbf{q}_{\underline{\mathbf{x}}',l} \tag{5.54}$$

and minimizing the previous expression with respect to the ς_l's, we find

$$\begin{aligned} \varsigma_l &= \frac{\lambda_{\underline{\mathbf{x}}',l}\mathbf{q}_{\underline{\mathbf{x}}',l}^T\underline{\mathbf{b}}_1'}{\underline{\mathbf{b}}_1'^T \mathbf{R}_{\underline{\mathbf{x}}'}\underline{\mathbf{b}}_1'} \\ &= \frac{\lambda_{\underline{\mathbf{x}}',l}\mathbf{q}_{\underline{\mathbf{x}}',l}^T\underline{\mathbf{b}}_1'}{\lambda_1'}, \quad l = 1, 2, \ldots, L, \end{aligned} \tag{5.55}$$

where $\lambda_1' = \underline{\mathbf{b}}_1'^T \mathbf{R}_{\underline{\mathbf{x}}'}\underline{\mathbf{b}}_1'$. Plugging these optimal values in (5.49), we obtain the optimal maximum SNR filtering matrix with minimum desired signal distortion:

$$\underline{\mathbf{H}}'_{\max} = \mathbf{\Lambda}_{\mathbf{x}'}\mathbf{T}_{\underline{\mathbf{x}}'}^T \frac{\underline{\mathbf{b}}_1'\underline{\mathbf{b}}_1'^T}{\lambda_1'}. \tag{5.56}$$

We also deduce that the maximum SNR filtering matrix for the estimation of $\mathbf{x}_1(k)$ is

$$\begin{aligned} \underline{\mathbf{H}}_{\max} &= \underline{\mathbf{I}}_{10}\mathbf{T}_{\underline{\mathbf{x}}'}\underline{\mathbf{H}}'_{\max} \\ &= \underline{\mathbf{I}}_{10}\mathbf{R}_{\underline{\mathbf{x}}'}\frac{\underline{\mathbf{b}}_1'\underline{\mathbf{b}}_1'^T}{\lambda_1'} \\ &= \underline{\mathbf{I}}_{10}\mathbf{T}_{\underline{\mathbf{x}}'}\mathbf{\Lambda}_{\mathbf{x}'}\mathbf{T}_{\underline{\mathbf{x}}'}^T\frac{\underline{\mathbf{b}}_1'\underline{\mathbf{b}}_1'^T}{\lambda_1'} \\ &= \underline{\mathbf{I}}_{10}\mathbf{R}_{\text{in}}\underline{\mathbf{b}}_1'\underline{\mathbf{b}}_1'^T. \end{aligned} \tag{5.57}$$

5.4.2 Wiener

By differentiating the MSE criterion, $J\left(\underline{\mathbf{H}}'\right)$, with respect to $\underline{\mathbf{H}}'$ and setting the result to zero, we find that the Wiener filtering matrix is

$$\begin{aligned}\underline{\mathbf{H}}'_{\mathrm{W}} &= \mathbf{\Lambda}_{\mathbf{x}'}\mathbf{T}_{\underline{\mathbf{x}}'}^T\mathbf{R}_{\underline{\mathbf{y}}}^{-1} \\ &= \mathbf{T}_{\underline{\mathbf{x}}'}^T\mathbf{R}_{\underline{\mathbf{x}}'}\mathbf{R}_{\underline{\mathbf{y}}}^{-1}.\end{aligned} \tag{5.58}$$

We deduce that the equivalent Wiener filtering matrix for the estimation of the vector $\mathbf{x}_1(k)$ is

$$\begin{aligned}\underline{\mathbf{H}}_{\mathrm{W}} &= \underline{\mathbf{I}}_{10}\mathbf{T}_{\underline{\mathbf{x}}'}\underline{\mathbf{H}}'_{\mathrm{W}} \\ &= \underline{\mathbf{I}}_{10}\mathbf{T}_{\underline{\mathbf{x}}'}\mathbf{\Lambda}_{\mathbf{x}'}\mathbf{T}_{\underline{\mathbf{x}}'}^T\mathbf{R}_{\underline{\mathbf{y}}}^{-1} \\ &= \underline{\mathbf{I}}_{10}\mathbf{R}_{\underline{\mathbf{x}}'}\mathbf{R}_{\underline{\mathbf{y}}'}^{-1} \\ &= \underline{\mathbf{I}}_{10}\mathbf{T}_{\underline{\mathbf{x}}'}\mathbf{\Lambda}_{\mathbf{x}'}\mathbf{T}_{\underline{\mathbf{x}}'}^T\left(\mathbf{T}_{\underline{\mathbf{x}}'}\mathbf{\Lambda}_{\mathbf{x}'}\mathbf{T}_{\underline{\mathbf{x}}'}^T + \mathbf{R}_{\mathrm{in}}\right)^{-1}.\end{aligned} \tag{5.59}$$

By applying the Woodbury's identity in (5.19) and then substituting the result in (5.58), we easily deduce another form of the Wiener filtering matrix:

$$\begin{aligned}\underline{\mathbf{H}}'_{\mathrm{W}} &= \left(\mathbf{I}_L + \mathbf{\Lambda}_{\mathbf{x}'}\mathbf{T}_{\underline{\mathbf{x}}'}^T\mathbf{R}_{\mathrm{in}}^{-1}\mathbf{T}_{\underline{\mathbf{x}}'}\right)^{-1}\mathbf{\Lambda}_{\mathbf{x}'}\mathbf{T}_{\underline{\mathbf{x}}'}^T\mathbf{R}_{\mathrm{in}}^{-1} \\ &= \left(\mathbf{\Lambda}_{\mathbf{x}'}^{-1} + \mathbf{T}_{\underline{\mathbf{x}}'}^T\mathbf{R}_{\mathrm{in}}^{-1}\mathbf{T}_{\underline{\mathbf{x}}'}\right)^{-1}\mathbf{T}_{\underline{\mathbf{x}}'}^T\mathbf{R}_{\mathrm{in}}^{-1}.\end{aligned} \tag{5.60}$$

The expression is interesting because it shows an obvious link with some other optimal rectangular filtering matrices as it will be verified later. We also have

$$\underline{\mathbf{H}}_{\mathrm{W}} = \underline{\mathbf{I}}_{10}\mathbf{T}_{\underline{\mathbf{x}}'}\left(\mathbf{I}_L + \mathbf{\Lambda}_{\mathbf{x}'}\mathbf{T}_{\underline{\mathbf{x}}'}^T\mathbf{R}_{\mathrm{in}}^{-1}\mathbf{T}_{\underline{\mathbf{x}}'}\right)^{-1}\mathbf{\Lambda}_{\mathbf{x}'}\mathbf{T}_{\underline{\mathbf{x}}'}^T\mathbf{R}_{\mathrm{in}}^{-1}. \tag{5.61}$$

Property 5.1. *The output SNR with the multichannel Wiener filtering matrix is always greater than or equal to the input SNR, i.e.,* $\mathrm{oSNR}\left(\underline{\mathbf{H}}_{\mathrm{W}}\right) \geq \mathrm{iSNR}$.

We also have

$$\mathrm{oSNR}\left(\underline{\mathbf{H}}_{\mathrm{W}}\right) \leq \mathrm{oSNR}\left(\underline{\mathbf{H}}_{\max}\right) \tag{5.62}$$

and, in general,

$$\upsilon_{\mathrm{sd}}\left(\underline{\mathbf{H}}_{\mathrm{W}}\right) \leq \upsilon_{\mathrm{sd}}\left(\underline{\mathbf{H}}_{\max}\right). \tag{5.63}$$

5.4.3 MVDR

The solution to the optimization problem:

$$\min_{\underline{\mathbf{H}}'} \mathrm{tr}\left(\underline{\mathbf{H}}'\mathbf{R}_{\mathrm{in}}\underline{\mathbf{H}}'^T\right) \quad \text{subject to} \quad \underline{\mathbf{H}}'\mathbf{T}_{\underline{\mathbf{x}}'} = \mathbf{I}_L \tag{5.64}$$

leads to the MVDR filtering matrix:

$$\underline{\mathbf{H}}'_{\mathrm{MVDR}} = \left(\mathbf{T}_{\underline{\mathbf{x}}'}^T\mathbf{R}_{\mathrm{in}}^{-1}\mathbf{T}_{\underline{\mathbf{x}}'}\right)^{-1}\mathbf{T}_{\underline{\mathbf{x}}'}^T\mathbf{R}_{\mathrm{in}}^{-1}, \tag{5.65}$$

which is interesting to compare to $\underline{\mathbf{H}}'_{\mathrm{W}}$ [eq. (5.60)]. Alternatively, we can rewrite the MVDR filtering matrix as

$$\underline{\mathbf{H}}'_{\mathrm{MVDR}} = \left(\mathbf{T}_{\underline{\mathbf{x}}'}^T\mathbf{R}_{\underline{\mathbf{y}}}^{-1}\mathbf{T}_{\underline{\mathbf{x}}'}\right)^{-1}\mathbf{T}_{\underline{\mathbf{x}}'}^T\mathbf{R}_{\underline{\mathbf{y}}}^{-1}. \tag{5.66}$$

From (5.66), we deduce the relationship between the MVDR and Wiener filtering matrices:

$$\underline{\mathbf{H}}'_{\mathrm{MVDR}} = \left(\underline{\mathbf{H}}'_{\mathrm{W}}\mathbf{T}_{\underline{\mathbf{x}}'}\right)^{-1}\underline{\mathbf{H}}'_{\mathrm{W}}. \tag{5.67}$$

We deduce that the MVDR filtering matrix for the estimation of $\mathbf{x}_1(k)$ is

$$\begin{aligned}\underline{\mathbf{H}}_{\mathrm{MVDR}} &= \underline{\mathbf{I}}_{10}\mathbf{T}_{\underline{\mathbf{x}}'}\left(\mathbf{T}_{\underline{\mathbf{x}}'}^T\mathbf{R}_{\mathrm{in}}^{-1}\mathbf{T}_{\underline{\mathbf{x}}'}\right)^{-1}\mathbf{T}_{\underline{\mathbf{x}}'}^T\mathbf{R}_{\mathrm{in}}^{-1} \\ &= \underline{\mathbf{I}}_{10}\mathbf{T}_{\underline{\mathbf{x}}'}\left(\mathbf{T}_{\underline{\mathbf{x}}'}^T\mathbf{R}_{\underline{\mathbf{y}}}^{-1}\mathbf{T}_{\underline{\mathbf{x}}'}\right)^{-1}\mathbf{T}_{\underline{\mathbf{x}}'}^T\mathbf{R}_{\underline{\mathbf{y}}}^{-1}.\end{aligned} \tag{5.68}$$

Property 5.2. *The output SNR with the multichannel MVDR filtering matrix is always greater than or equal to the input SNR, i.e.,* $\mathrm{oSNR}\left(\underline{\mathbf{H}}_{\mathrm{MVDR}}\right) \geq \mathrm{iSNR}$.

We also have

$$\mathrm{oSNR}\left(\underline{\mathbf{H}}_{\mathrm{MVDR}}\right) \leq \mathrm{oSNR}\left(\underline{\mathbf{H}}_{\mathrm{W}}\right) \leq \mathrm{oSNR}\left(\underline{\mathbf{H}}_{\max}\right) \tag{5.69}$$

and

$$\xi_{\mathrm{sr}}\left(\underline{\mathbf{H}}_{\mathrm{MVDR}}\right) = 1, \tag{5.70}$$

$$\upsilon_{\mathrm{sd}}\left(\underline{\mathbf{H}}_{\mathrm{MVDR}}\right) = 0. \tag{5.71}$$

5.4.4 Tradeoff

In the tradeoff approach, we optimize the following criterion:

$$\min_{\underline{\mathbf{H}}'} J_{\text{ds}}\left(\underline{\mathbf{H}}'\right) \quad \text{subject to} \quad J_{\text{rs}}\left(\underline{\mathbf{H}}'\right) = \beta \text{tr}\left(\mathbf{R}_{\underline{\mathbf{v}}}\right), \tag{5.72}$$

where $0 < \beta < 1$ to insure that we get some noise reduction. We easily find that the optimal tradeoff filtering matrix is

$$\underline{\mathbf{H}}'_{\text{T},\mu} = \mathbf{\Lambda}_{\mathbf{x}'}\mathbf{T}^T_{\underline{\mathbf{x}}'}\left(\mathbf{T}_{\underline{\mathbf{x}}'}\mathbf{\Lambda}_{\mathbf{x}'}\mathbf{T}^H_{\underline{\mathbf{x}}'} + \mu\mathbf{R}_{\text{in}}\right)^{-1}, \tag{5.73}$$

where $\mu > 0$ is a Lagrange multiplier. We can express (5.73) as

$$\underline{\mathbf{H}}'_{\text{T},\mu} = \left(\mu\mathbf{\Lambda}^{-1}_{\mathbf{x}'} + \mathbf{T}^T_{\underline{\mathbf{x}}'}\mathbf{R}^{-1}_{\text{in}}\mathbf{T}_{\underline{\mathbf{x}}'}\right)^{-1}\mathbf{T}^T_{\underline{\mathbf{x}}'}\mathbf{R}^{-1}_{\text{in}}. \tag{5.74}$$

We observe from the previous equation that for

- $\mu = 1, \underline{\mathbf{H}}'_{\text{T},1} = \underline{\mathbf{H}}'_{\text{W}}$, which is the Wiener filtering matrix;
- $\mu = 0, \underline{\mathbf{H}}'_{\text{T},0} = \underline{\mathbf{H}}'_{\text{MVDR}}$, which is the MVDR filtering matrix;
- $\mu > 1$, results in a filtering matrix with low residual noise at the expense of high speech distortion (as compared to Wiener);
- $\mu < 1$, results in a filtering matrix with high residual noise and low speech distortion (as compared to Wiener).

We deduce that the tradeoff filtering matrix for the estimation of $\mathbf{x}_1(k)$ is

$$\underline{\mathbf{H}}_{\text{T},\mu} = \underline{\mathbf{I}}_{10}\mathbf{T}_{\underline{\mathbf{x}}'}\left(\mu\mathbf{\Lambda}^{-1}_{\mathbf{x}'} + \mathbf{T}^T_{\underline{\mathbf{x}}'}\mathbf{R}^{-1}_{\text{in}}\mathbf{T}_{\underline{\mathbf{x}}'}\right)^{-1}\mathbf{T}^T_{\underline{\mathbf{x}}'}\mathbf{R}^{-1}_{\text{in}}, \tag{5.75}$$

which clearly shows how the speech subspace should be modified in order to make a compromise between noise reduction and speech distortion.

Property 5.3. *The output SNR with the multichannel tradeoff filtering matrix is always greater than or equal to the input SNR, i.e.,* $\text{oSNR}\left(\underline{\mathbf{H}}_{\text{T},\mu}\right) \geq \text{iSNR}, \forall\mu \geq 0$.

We should have for $\mu \geq 1$,

$$\text{oSNR}\left(\underline{\mathbf{H}}_{\text{MVDR}}\right) \leq \text{oSNR}\left(\underline{\mathbf{H}}_{\text{W}}\right) \leq \text{oSNR}\left(\underline{\mathbf{H}}_{\text{T},\mu}\right) \leq \text{oSNR}\left(\underline{\mathbf{H}}_{\text{max}}\right), \tag{5.76}$$

$$0 = \upsilon_{\text{sd}}\left(\underline{\mathbf{H}}_{\text{MVDR}}\right) \leq \upsilon_{\text{sd}}\left(\underline{\mathbf{H}}_{\text{W}}\right) \leq \upsilon_{\text{sd}}\left(\underline{\mathbf{H}}_{\text{T},\mu}\right), \tag{5.77}$$

and for $\mu \leq 1$,

$$\text{oSNR}\left(\underline{\mathbf{H}}_{\text{MVDR}}\right) \leq \text{oSNR}\left(\underline{\mathbf{H}}_{\text{T},\mu}\right) \leq \text{oSNR}\left(\underline{\mathbf{H}}_{\text{W}}\right) \leq \text{oSNR}\left(\underline{\mathbf{H}}_{\max}\right), \tag{5.78}$$

$$0 = \upsilon_{\text{sd}}\left(\underline{\mathbf{H}}_{\text{MVDR}}\right) \leq \upsilon_{\text{sd}}\left(\underline{\mathbf{H}}_{\text{T},\mu}\right) \leq \upsilon_{\text{sd}}\left(\underline{\mathbf{H}}_{\text{W}}\right). \tag{5.79}$$

5.4.5 LCMV

Let us decompose the noise signal vector, $\underline{\mathbf{v}}(k)$, into two orthogonal components, i.e.,

$$\begin{aligned}\underline{\mathbf{v}}(k) &= \mathbf{R}_{\underline{\mathbf{v}}\mathbf{v}_1}\mathbf{R}_{\mathbf{v}_1}^{-1}\mathbf{v}_1(k) + \underline{\mathbf{v}}_{\text{u}}(k) \\ &= \boldsymbol{\Gamma}_{\underline{\mathbf{v}}\mathbf{v}_1}\mathbf{v}_1(k) + \underline{\mathbf{v}}_{\text{u}}(k) \\ &= \underline{\mathbf{v}}'(k) + \underline{\mathbf{v}}_{\text{u}}(k),\end{aligned} \tag{5.80}$$

where $\underline{\mathbf{v}}'(k)$ and $\underline{\mathbf{v}}_{\text{u}}(k)$ are the correlated and uncorrelated noise signal vectors [with respect to $\mathbf{v}_1(k)$], respectively, and

$$E\left[\underline{\mathbf{v}}'(k)\underline{\mathbf{v}}_{\text{u}}^T(k)\right] = \mathbf{0}_{ML\times ML}. \tag{5.81}$$

Using the decomposition:

$$\begin{aligned}\mathbf{R}_{\underline{\mathbf{v}}'} &= \mathbf{Q}_{\underline{\mathbf{v}}'}\boldsymbol{\Lambda}_{\underline{\mathbf{v}}'}\mathbf{Q}_{\underline{\mathbf{v}}'}^T \\ &= \left[\begin{array}{cc}\mathbf{T}_{\underline{\mathbf{v}}'} & \boldsymbol{\Xi}_{\underline{\mathbf{v}}'}\end{array}\right]\boldsymbol{\Lambda}_{\underline{\mathbf{v}}'}\left[\begin{array}{c}\mathbf{T}_{\underline{\mathbf{v}}'}^T \\ \boldsymbol{\Xi}_{\underline{\mathbf{v}}'}^T\end{array}\right],\end{aligned} \tag{5.82}$$

where all the matrices involved in (5.82) are defined in a similar way to the matrices involved in (5.12), we can rewrite (5.80) as

$$\underline{\mathbf{v}}(k) = \mathbf{T}_{\underline{\mathbf{v}}'}\mathbf{T}_{\underline{\mathbf{v}}'}^T\underline{\mathbf{v}}'(k) + \underline{\mathbf{v}}_{\text{u}}(k). \tag{5.83}$$

The LCMV approach consists of estimating $\underline{\mathbf{x}}'(k)$ without any distortion, completely removing the correlated noise, and attenuating the uncorrelated noise as much as possible. It follows that the constraints are

$$\underline{\mathbf{H}}'\mathbf{C}_{\underline{\mathbf{x}}'\underline{\mathbf{v}}'} = \left[\begin{array}{cc}\mathbf{I}_L & \mathbf{0}_{L\times L}\end{array}\right], \tag{5.84}$$

where

$$\mathbf{C}_{\underline{\mathbf{x}}'\underline{\mathbf{v}}'} = \left[\begin{array}{cc}\mathbf{T}_{\underline{\mathbf{x}}'} & \mathbf{T}_{\underline{\mathbf{v}}'}\end{array}\right] \tag{5.85}$$

is the constraint matrix of size $ML \times 2L$. Our optimization problem is now

$$\min_{\underline{\mathbf{H}}'} \operatorname{tr}\left(\underline{\mathbf{H}}'\mathbf{R}_{\underline{\mathbf{y}}}\underline{\mathbf{H}}'^{T}\right) \quad \text{subject to} \quad \underline{\mathbf{H}}'\mathbf{C}_{\underline{\mathbf{x}}'\underline{\mathbf{y}}'} = \left[\, \mathbf{I}_L \; \mathbf{0}_{L\times L} \,\right], \tag{5.86}$$

from which we find the LCMV filtering matrix:

$$\underline{\mathbf{H}}'_{\text{LCMV}} = \left[\, \mathbf{I}_L \; \mathbf{0}_{L\times L} \,\right]\left(\mathbf{C}^{T}_{\underline{\mathbf{x}}'\underline{\mathbf{y}}'}\mathbf{R}^{-1}_{\underline{\mathbf{y}}}\mathbf{C}_{\underline{\mathbf{x}}'\underline{\mathbf{y}}'}\right)^{-1}\mathbf{C}^{T}_{\underline{\mathbf{x}}'\underline{\mathbf{y}}'}\mathbf{R}^{-1}_{\underline{\mathbf{y}}}. \tag{5.87}$$

We deduce that the LCMV for the estimation of $\mathbf{x}_1(k)$ is

$$\underline{\mathbf{H}}_{\text{LCMV}} = \underline{\mathbf{I}}_{10}\mathbf{T}_{\underline{\mathbf{x}}'}\left[\, \mathbf{I}_L \; \mathbf{0}_{L\times L} \,\right]\left(\mathbf{C}^{T}_{\underline{\mathbf{x}}'\underline{\mathbf{y}}'}\mathbf{R}^{-1}_{\underline{\mathbf{y}}}\mathbf{C}_{\underline{\mathbf{x}}'\underline{\mathbf{y}}'}\right)^{-1}\mathbf{C}^{T}_{\underline{\mathbf{x}}'\underline{\mathbf{y}}'}\mathbf{R}^{-1}_{\underline{\mathbf{y}}}. \tag{5.88}$$

REFERENCES

[1] J. Benesty, J. Chen, Y. Huang, Microphone Array Signal Processing, Springer-Verlag, Berlin, Germany, 2008.

[2] M. Brandstein, D.B. Ward (Eds.),, Microphone Arrays: Signal Processing Techniques and Applications, Springer-Verlag, Berlin, Germany, 2001.

[3] J. Benesty, J. Chen, Optimal Time-domain Noise Reduction Filters–A Theoretical Study, Springer Briefs in Electrical and Computer Engineering, Springer-Verlag, 2011.

CHAPTER 6

Multichannel Speech Enhancement in the Frequency Domain

In this chapter, we study the multichannel speech enhancement problem in the frequency domain. By exploiting the structure of the speech subspace, we can easily estimate all convolved speech signals at the microphones with a simple complex filter. As a result, binaural noise reduction with this approach is straightforward since we can choose any two signals from the estimated convolved speech signals, which contain all the necessary spatial information for the localization of the desired source signal.

6.1 SIGNAL MODEL AND PROBLEM FORMULATION

We consider a sensor array consisting of M microphones. In a general way, the received signals at the frequency index f are expressed as [1,2]

$$\begin{aligned} Y_m(f) &= G_m(f)S(f) + V_m(f) \\ &= X_m(f) + V_m(f), \quad m = 1, 2, \ldots, M, \end{aligned} \tag{6.1}$$

where $Y_m(f)$ is the mth microphone signal, $S(f)$ is the unknown speech source, $G_m(f)$ is the acoustic impulse response from the position of $S(f)$ to the mth microphone, $X_m(f) = G_m(f)S(f)$ is the convolved speech signal, and $V_m(f)$ is the additive noise. It is assumed that $X_m(f)$ and $V_m(f)$ are incoherent and zero mean.

It is more convenient to write the M frequency-domain microphone signals in a vector notation:

$$\begin{aligned} \mathbf{y}(f) &= \mathbf{g}(f)S(f) + \mathbf{v}(f) \\ &= \mathbf{x}(f) + \mathbf{v}(f), \end{aligned} \tag{6.2}$$

where

$$\begin{aligned} \mathbf{y}(f) &= \left[Y_1(f)\ Y_2(f)\ \cdots\ Y_M(f) \right]^T, \\ \mathbf{x}(f) &= \left[X_1(f)\ X_2(f)\ \cdots\ X_M(f) \right]^T \\ &= \mathbf{g}(f)S(f), \end{aligned}$$

Speech Enhancement. http://dx.doi.org/10.1016/B978-0-12-800139-4.00006-2

$$\mathbf{g}(f) = \left[G_1(f)\ G_2(f) \cdots G_M(f) \right]^T,$$
$$\mathbf{v}(f) = \left[V_1(f)\ V_2(f) \cdots V_M(f) \right]^T.$$

In the rest, we assume that the whole vector $\mathbf{x}(f)$ is the desired signal that we wish to estimate from the observation vector, $\mathbf{y}(f)$. Since $S(f)$ and $V_m(f)$ are incoherent by assumption, the correlation matrix of $\mathbf{y}(f)$ is

$$\begin{aligned}\mathbf{\Phi}_{\mathbf{y}}(f) &= E\left[\mathbf{y}(f)\mathbf{y}^H(f)\right] \\ &= \mathbf{\Phi}_{\mathbf{x}}(f) + \mathbf{\Phi}_{\mathbf{v}}(f) \\ &= \phi_S(f)\mathbf{g}(f)\mathbf{g}^H(f) + \mathbf{\Phi}_{\mathbf{v}}(f), \end{aligned} \tag{6.3}$$

where $\mathbf{\Phi}_{\mathbf{x}}(f) = \phi_S(f)\mathbf{g}(f)\mathbf{g}^H(f)$ is the correlation matrix (whose rank is equal to 1) of $\mathbf{x}(f)$, $\phi_S(f) = E\left[|S(f)|^2\right]$ is the variance of $S(f)$, and $\mathbf{\Phi}_{\mathbf{v}}(f) = E\left[\mathbf{v}(f)\mathbf{v}^H(f)\right]$ is the correlation matrix (whose rank is assumed to be equal to M) of $\mathbf{v}(f)$.

The speech spatial correlation matrix can be decomposed as follows:

$$\mathbf{\Phi}_{\mathbf{x}}(f) = \mathbf{Q}_{\mathbf{x}}(f)\mathbf{\Lambda}_{\mathbf{x}}(f)\mathbf{Q}_{\mathbf{x}}^H(f), \tag{6.4}$$

where

$$\mathbf{Q}_{\mathbf{x}}(f) = \left[\mathbf{q}_{\mathbf{x},1}(f)\ \mathbf{q}_{\mathbf{x},2}(f) \cdots \mathbf{q}_{\mathbf{x},M}(f) \right] \tag{6.5}$$

is a unitary matrix and

$$\mathbf{\Lambda}_{\mathbf{x}}(f) = \text{diag}\left[\lambda_{\mathbf{x},1}(f), \lambda_{\mathbf{x},2}(f), \ldots, \lambda_{\mathbf{x},M}(f)\right] \tag{6.6}$$

is a diagonal matrix. The orthonormal vectors $\mathbf{q}_{\mathbf{x},1}(f), \mathbf{q}_{\mathbf{x},2}(f), \ldots, \mathbf{q}_{\mathbf{x},M}(f)$ are the eigenvectors corresponding, respectively, to the eigenvalues $\lambda_{\mathbf{x},1}(f), \lambda_{\mathbf{x},2}(f), \ldots, \lambda_{\mathbf{x},M}(f)$ of the matrix $\mathbf{\Phi}_{\mathbf{x}}(f)$, where

$$\begin{aligned}\lambda_{\mathbf{x},1}(f) &= \text{tr}\left[\mathbf{\Phi}_{\mathbf{x}}(f)\right] \\ &= \phi_S(f)\mathbf{g}^H(f)\mathbf{g}(f) \end{aligned} \tag{6.7}$$

and $\lambda_{\mathbf{x},2}(f) = \lambda_{\mathbf{x},3}(f) = \cdots = \lambda_{\mathbf{x},M}(f) = 0$. Let

$$\mathbf{Q}_{\mathbf{x}}(f) = \left[\mathbf{q}_{\mathbf{x},1}(f)\ \mathbf{\Upsilon}_{\mathbf{x}}(f) \right], \tag{6.8}$$

where

$$\mathbf{q}_{\mathbf{x},1}(f) = \frac{\mathbf{g}(f)}{\sqrt{\mathbf{g}^H(f)\mathbf{g}(f)}} \tag{6.9}$$

and

$$\boldsymbol{\Upsilon}_{\mathbf{x}}(f) = \left[\mathbf{q}_{\mathbf{x},2}(f)\ \mathbf{q}_{\mathbf{x},3}(f) \cdots \mathbf{q}_{\mathbf{x},M}(f) \right]. \tag{6.10}$$

It can be verified that

$$\mathbf{I}_M = \mathbf{q}_{\mathbf{x},1}(f)\mathbf{q}_{\mathbf{x},1}^H(f) + \boldsymbol{\Upsilon}_{\mathbf{x}}(f)\boldsymbol{\Upsilon}_{\mathbf{x}}^H(f). \tag{6.11}$$

Notice that $\mathbf{q}_{\mathbf{x},1}(f)\mathbf{q}_{\mathbf{x},1}^H(f)$ and $\boldsymbol{\Upsilon}_{\mathbf{x}}(f)\boldsymbol{\Upsilon}_{\mathbf{x}}^H(f)$ are two orthogonal projection matrices of rank 1 and $M-1$, respectively. Hence, $\mathbf{q}_{\mathbf{x},1}(f)\mathbf{q}_{\mathbf{x},1}^H(f)$ is the orthogonal projector onto the speech subspace (where all the energy of the speech signal is concentrated) or range of $\boldsymbol{\Phi}_{\mathbf{x}}(f)$ and $\boldsymbol{\Upsilon}_{\mathbf{x}}(f)\boldsymbol{\Upsilon}_{\mathbf{x}}^H(f)$ is the orthogonal projector onto the null subspace of $\boldsymbol{\Phi}_{\mathbf{x}}(f)$. Using (6.11), we can write the speech vector as

$$\begin{aligned}
\mathbf{x}(f) &= \mathbf{Q}_{\mathbf{x}}(f)\mathbf{Q}_{\mathbf{x}}^H(f)\mathbf{x}(f) \\
&= \mathbf{q}_{\mathbf{x},1}(f)\mathbf{q}_{\mathbf{x},1}^H(f)\mathbf{x}(f) \\
&= \mathbf{q}_{\mathbf{x},1}(f)\widetilde{X}(f),
\end{aligned} \tag{6.12}$$

where

$$\widetilde{X}(f) = \mathbf{q}_{\mathbf{x},1}^H(f)\mathbf{x}(f) \tag{6.13}$$

is the transformed desired signal element. Therefore, the signal model for multichannel noise reduction becomes

$$\mathbf{y}(f) = \mathbf{q}_{\mathbf{x},1}(f)\widetilde{X}(f) + \mathbf{v}(f). \tag{6.14}$$

We see that the estimation of the vector $\mathbf{x}(f)$ of length M is equivalent to the estimation of the component $\widetilde{X}(f)$. From (6.14), we give another form of the correlation matrix of $\mathbf{y}(f)$:

$$\boldsymbol{\Phi}_{\mathbf{y}}(f) = \phi_{\widetilde{X}}(f)\mathbf{q}_{\mathbf{x},1}(f)\mathbf{q}_{\mathbf{x},1}^H(f) + \boldsymbol{\Phi}_{\mathbf{v}}(f), \tag{6.15}$$

where $\phi_{\widetilde{X}}(f) = E\left[\left|\widetilde{X}(f)\right|^2\right] = \lambda_{\mathbf{x},1}(f)$ and, obviously, $\boldsymbol{\Phi}_{\mathbf{x}}(f) = \lambda_{\mathbf{x},1}(f)$ $\mathbf{q}_{\mathbf{x},1}(f)\mathbf{q}_{\mathbf{x},1}^H(f)$.

6.2 LINEAR ARRAY MODEL

In our context, multichannel noise reduction consists of estimating $\widetilde{X}(f)$ from the observations. This task is performed in the same way as the classical beamforming approach, i.e., by applying a complex weight to the output of each sensor, at frequency f, and summing across the aperture [3–5]:

$$\begin{aligned}\widetilde{Z}(f) &= \sum_{m=1}^{M} \widetilde{H}_m^*(f) Y_m(f) \\ &= \widetilde{\mathbf{h}}^H(f)\mathbf{y}(f),\end{aligned} \tag{6.16}$$

where $\widetilde{Z}(f)$ is supposed to be the estimate of $\widetilde{X}(f)$ and

$$\widetilde{\mathbf{h}}(f) = \left[\widetilde{H}_1(f)\ \widetilde{H}_2(f) \cdots \widetilde{H}_M(f) \right]^T \tag{6.17}$$

is a complex-valued filter of length M. We can rewrite (6.16) as

$$\widetilde{Z}(f) = \widetilde{X}_{\rm fd}(f) + \widetilde{V}_{\rm rn}(f), \tag{6.18}$$

where

$$\begin{aligned}\widetilde{X}_{\rm fd}(f) &= \widetilde{\mathbf{h}}^H(f)\mathbf{x}(f) \\ &= \widetilde{\mathbf{h}}^H(f)\mathbf{q}_{\mathbf{x},1}(f)\widetilde{X}(f)\end{aligned} \tag{6.19}$$

is the filtered transformed desired signal and

$$\widetilde{V}_{\rm rn}(f) = \widetilde{\mathbf{h}}^H(f)\mathbf{v}(f) \tag{6.20}$$

is the residual noise. Equivalently, the estimate of $\mathbf{x}(f)$ is supposed to be

$$\begin{aligned}\mathbf{z}(f) &= \mathbf{q}_{\mathbf{x},1}(f)\widetilde{Z}(f) \\ &= \mathbf{q}_{\mathbf{x},1}(f)\widetilde{\mathbf{h}}^H(f)\mathbf{y}(f) \\ &= \mathbf{H}(f)\mathbf{y}(f),\end{aligned} \tag{6.21}$$

where

$$\mathbf{H}(f) = \mathbf{q}_{\mathbf{x},1}(f)\widetilde{\mathbf{h}}^H(f) \tag{6.22}$$

is a filtering matrix of size $M \times M$ that leads to the estimation of $\mathbf{x}(f)$. Now, it is easy to compute the variance of $\widetilde{Z}(f)$, which is

$$\begin{aligned}\phi_{\widetilde{Z}}(f) &= E\left[\left|\widetilde{Z}(f)\right|^2\right] \\ &= \phi_{\widetilde{X}_{\rm fd}}(f) + \phi_{\widetilde{V}_{\rm rn}}(f),\end{aligned} \tag{6.23}$$

where

$$\begin{aligned}\phi_{\widetilde{X}_{\rm fd}}(f) &= \widetilde{\mathbf{h}}^H(f)\mathbf{\Phi}_{\mathbf{x}}(f)\widetilde{\mathbf{h}}(f) \\ &= \phi_{\widetilde{X}}(f)\left|\widetilde{\mathbf{h}}^H(f)\mathbf{q}_{\mathbf{x},1}(f)\right|^2,\end{aligned} \tag{6.24}$$

$$\phi_{\widetilde{V}_{\rm rn}}(f) = \widetilde{\mathbf{h}}^H(f)\mathbf{\Phi}_{\mathbf{v}}(f)\widetilde{\mathbf{h}}(f). \tag{6.25}$$

We also observe that $\mathbf{\Phi}_{\mathbf{z}}(f) = \phi_{\widetilde{Z}}(f)\mathbf{q}_{\mathbf{x},1}(f)\mathbf{q}_{\mathbf{x},1}^{H}(f)$ and $\mathrm{tr}\left[\mathbf{\Phi}_{\mathbf{z}}(f)\right] = \phi_{\widetilde{Z}}(f)$. The variance of $\widetilde{Z}(f)$ is helpful in defining meaningful performance measures.

6.3 PERFORMANCE MEASURES

In this section, we define some important performance measures for multichannel noise reduction in the frequency domain. We discuss both narrowband and broadband measures.

6.3.1 Noise Reduction

Since $\widetilde{X}(f)$ is the transformed desired signal, the narrowband input SNR is

$$\begin{aligned} \mathrm{iSNR}(f) &= \frac{\mathrm{tr}\left[\mathbf{\Phi}_{\mathbf{x}}(f)\right]}{\mathrm{tr}\left[\mathbf{\Phi}_{\mathbf{v}}(f)\right]} \\ &= \frac{\phi_{\widetilde{X}}(f)}{\mathrm{tr}\left[\mathbf{\Phi}_{\mathbf{v}}(f)\right]}. \end{aligned} \tag{6.26}$$

From (6.26), we deduce the broadband input SNR:

$$\mathrm{iSNR} = \frac{\int_f \phi_{\widetilde{X}}(f)df}{\int_f \mathrm{tr}\left[\mathbf{\Phi}_{\mathbf{v}}(f)\right]df}. \tag{6.27}$$

It can be shown that

$$\min_m \frac{\phi_{X_m}(f)}{\phi_{V_m}(f)} \leq \mathrm{iSNR}(f) \leq \max_m \frac{\phi_{X_m}(f)}{\phi_{V_m}(f)}, \tag{6.28}$$

where $\phi_{X_m}(f)$ and $\phi_{V_m}(f)$ are the variances of $X_m(f)$ and $V_m(f)$, respectively.

From (6.23), we find the narrowband output SNR:

$$\begin{aligned} \mathrm{oSNR}\left[\widetilde{\mathbf{h}}(f)\right] &= \frac{\phi_{\widetilde{X}_{\mathrm{fd}}}(f)}{\phi_{\widetilde{V}_{\mathrm{rn}}}(f)} \\ &= \frac{\phi_{\widetilde{X}}(f)\left|\widetilde{\mathbf{h}}^{H}(f)\mathbf{q}_{\mathbf{x},1}(f)\right|^2}{\widetilde{\mathbf{h}}^{H}(f)\mathbf{\Phi}_{\mathbf{v}}(f)\widetilde{\mathbf{h}}(f)} \\ &= \frac{1}{|G_m|^2} \times \frac{\phi_{X_m}(f)\left|\widetilde{\mathbf{h}}^{H}(f)\mathbf{g}(f)\right|^2}{\widetilde{\mathbf{h}}^{H}(f)\mathbf{\Phi}_{\mathbf{v}}(f)\widetilde{\mathbf{h}}(f)} \end{aligned} \tag{6.29}$$

and the broadband output SNR:

$$\text{oSNR}\left(\widetilde{\mathbf{h}}\right) = \frac{\int_f \phi_{\widetilde{X}}(f) \left|\widetilde{\mathbf{h}}^H(f)\mathbf{q}_{\mathbf{x},1}(f)\right|^2 df}{\int_f \widetilde{\mathbf{h}}^H(f)\mathbf{\Phi}_{\mathbf{v}}(f)\widetilde{\mathbf{h}}(f)df}. \tag{6.30}$$

Since it is assumed that the matrix $\mathbf{\Phi}_{\mathbf{v}}(f)$ is full rank, we have

$$\left|\widetilde{\mathbf{h}}^H(f)\mathbf{q}_{\mathbf{x},1}(f)\right|^2 \leq \left[\widetilde{\mathbf{h}}^H(f)\mathbf{\Phi}_{\mathbf{v}}(f)\widetilde{\mathbf{h}}(f)\right]\left[\mathbf{q}_{\mathbf{x},1}^H(f)\mathbf{\Phi}_{\mathbf{v}}^{-1}(f)\mathbf{q}_{\mathbf{x},1}(f)\right], \tag{6.31}$$

with equality if and only if $\widetilde{\mathbf{h}}(f) \propto \mathbf{\Phi}_{\mathbf{v}}^{-1}(f)\mathbf{q}_{\mathbf{x},1}(f)$. Using the inequality (6.31) in (6.29), we find an upper bound for the narrowband output SNR:

$$\begin{aligned}\text{oSNR}\left[\widetilde{\mathbf{h}}(f)\right] &\leq \phi_{\widetilde{X}}(f)\mathbf{q}_{\mathbf{x},1}^H(f)\mathbf{\Phi}_{\mathbf{v}}^{-1}(f)\mathbf{q}_{\mathbf{x},1}(f) \\ &= \frac{\phi_{X_m}(f)}{|G_m|^2}\mathbf{g}^H(f)\mathbf{\Phi}_{\mathbf{v}}^{-1}(f)\mathbf{g}(f), \forall\widetilde{\mathbf{h}}(f).\end{aligned} \tag{6.32}$$

As a consequence, we define the maximum narrowband output SNR as

$$\begin{aligned}\text{oSNR}_{\max}(f) &= \phi_{\widetilde{X}}(f)\mathbf{q}_{\mathbf{x},1}^H(f)\mathbf{\Phi}_{\mathbf{v}}^{-1}(f)\mathbf{q}_{\mathbf{x},1}(f) \\ &= \text{tr}\left[\mathbf{\Phi}_{\mathbf{v}}^{-1}(f)\mathbf{\Phi}_{\mathbf{x}}(f)\right].\end{aligned} \tag{6.33}$$

For the particular filters, $\mathbf{i}_m,\ m = 1, 2, \ldots, M$, where $\mathbf{i}_m$ corresponds to the mth column of the identity matrix $\mathbf{I}_M$, we have

$$\text{oSNR}\left[\mathbf{i}_m(f)\right] = \frac{\phi_{X_m}(f)}{\phi_{V_m}(f)}. \tag{6.34}$$

As a result,

$$\text{oSNR}_{\max}(f) \geq \max_m \frac{\phi_{X_m}(f)}{\phi_{V_m}(f)} \geq \text{iSNR}(f). \tag{6.35}$$

It follows from the definitions of the input and output SNRs that the narrowband and broadband array gains are, respectively,

$$\mathcal{A}\left[\widetilde{\mathbf{h}}(f)\right] = \frac{\text{oSNR}\left[\widetilde{\mathbf{h}}(f)\right]}{\text{iSNR}(f)}, \tag{6.36}$$

$$\mathcal{A}\left(\widetilde{\mathbf{h}}\right) = \frac{\text{oSNR}\left(\widetilde{\mathbf{h}}\right)}{\text{iSNR}}. \tag{6.37}$$

The noise reduction factor is defined as the power of the noise at the microphones over the power of the noise remaining after filtering. We have the narrowband noise reduction factor:

$$\xi_{\mathrm{nr}}\left[\widetilde{\mathbf{h}}(f)\right] = \frac{\mathrm{tr}\left[\mathbf{\Phi}_{\mathbf{v}}(f)\right]}{\widetilde{\mathbf{h}}^H(f)\mathbf{\Phi}_{\mathbf{v}}(f)\widetilde{\mathbf{h}}(f)} \tag{6.38}$$

and the broadband noise reduction factor:

$$\xi_{\mathrm{nr}}\left(\widetilde{\mathbf{h}}\right) = \frac{\int_f \mathrm{tr}\left[\mathbf{\Phi}_{\mathbf{v}}(f)\right] df}{\int_f \widetilde{\mathbf{h}}^H(f)\mathbf{\Phi}_{\mathbf{v}}(f)\widetilde{\mathbf{h}}(f)df}. \tag{6.39}$$

The noise reduction factor should be greater than 1 for optimal filters.

6.3.2 Speech Distortion

To measure the distortion of the transformed desired signal, we can use the narrowband speech reduction factor:

$$\begin{aligned} \xi_{\mathrm{sr}}\left[\widetilde{\mathbf{h}}(f)\right] &= \frac{\phi_{\widetilde{X}}(f)}{\phi_{\widetilde{X}_{\mathrm{fd}}}(f)} \\ &= \frac{1}{\left|\widetilde{\mathbf{h}}^H(f)\mathbf{q}_{\mathbf{x},1}(f)\right|^2} \end{aligned} \tag{6.40}$$

and the broadband speech reduction factor:

$$\xi_{\mathrm{sr}}\left(\widetilde{\mathbf{h}}\right) = \frac{\int_f \phi_{\widetilde{X}}(f)df}{\int_f \phi_{\widetilde{X}}(f)\left|\widetilde{\mathbf{h}}^H(f)\mathbf{q}_{\mathbf{x},1}(f)\right|^2 df}. \tag{6.41}$$

To avoid any distortion of the transformed desired signal, we must have

$$\widetilde{\mathbf{h}}^H(f)\mathbf{q}_{\mathbf{x},1}(f) = 1, \forall f. \tag{6.42}$$

So when the speech reduction factor is greater than 1, the transformed desired signal is distorted.

It is clear that we have the following relationships:

$$\mathcal{A}\left(\widetilde{\mathbf{h}}\right) = \frac{\xi_{\mathrm{nr}}\left(\widetilde{\mathbf{h}}\right)}{\xi_{\mathrm{sr}}\left(\widetilde{\mathbf{h}}\right)}, \tag{6.43}$$

$$\mathcal{A}\left[\widetilde{\mathbf{h}}(f)\right] = \frac{\xi_{\mathrm{nr}}\left[\widetilde{\mathbf{h}}(f)\right]}{\xi_{\mathrm{sr}}\left[\widetilde{\mathbf{h}}(f)\right]}. \tag{6.44}$$

We can also measure distortion via the narrowband speech distortion index:

$$\upsilon_{\text{sd}}\left[\widetilde{\mathbf{h}}(f)\right] = \frac{E\left[\left|\widetilde{X}_{\text{fd}}(f) - \widetilde{X}(f)\right|^2\right]}{\phi_{\widetilde{X}}(f)}$$
$$= \left|\widetilde{\mathbf{h}}^H(f)\mathbf{q}_{\mathbf{x},1}(f) - 1\right|^2. \quad (6.45)$$

We deduce that the broadband speech distortion index is

$$\upsilon_{\text{sd}}\left(\widetilde{\mathbf{h}}\right) = \frac{\int_f \phi_{\widetilde{X}}(f)\left|\widetilde{\mathbf{h}}^H(f)\mathbf{q}_{\mathbf{x},1}(f) - 1\right|^2 df}{\int_f \phi_{\widetilde{X}}(f)df}$$
$$= \frac{\int_f \phi_{\widetilde{X}}(f)\upsilon_{\text{sd}}\left[\widetilde{\mathbf{h}}(f)\right]df}{\int_f \phi_{\widetilde{X}}(f)df}. \quad (6.46)$$

The speech distortion index should be smaller than 1 for optimal filters and equal to 0 in the distortionless case.

6.3.3 MSE Criterion

We define the error signal between the estimated and transformed desired signals, $\widetilde{Z}(f)$ and $\widetilde{X}(f)$, at frequency f, as

$$\widetilde{\mathcal{E}}(f) = \widetilde{Z}(f) - \widetilde{X}(f)$$
$$= \widetilde{\mathbf{h}}^H(f)\mathbf{y}(f) - \widetilde{X}(f)$$
$$= \widetilde{X}_{\text{fd}}(f) + \widetilde{V}_{\text{rn}}(f) - \widetilde{X}(f). \quad (6.47)$$

This error can also be expressed as

$$\widetilde{\mathcal{E}}(f) = \widetilde{\mathcal{E}}_{\text{ds}}(f) + \widetilde{\mathcal{E}}_{\text{rs}}(f), \quad (6.48)$$

where

$$\widetilde{\mathcal{E}}_{\text{ds}}(f) = \left[\widetilde{\mathbf{h}}^H(f)\mathbf{q}_{\mathbf{x},1}(f) - 1\right]\widetilde{X}(f) \quad (6.49)$$

is the speech distortion due to the complex filter and

$$\widetilde{\mathcal{E}}_{\text{rs}}(f) = \widetilde{\mathbf{h}}^H(f)\mathbf{v}(f) \quad (6.50)$$

represents the residual noise. Notice that the error signals $\widetilde{\mathcal{E}}_{\text{ds}}(f)$ and $\widetilde{\mathcal{E}}_{\text{rs}}(f)$ are incoherent. The narrowband MSE is then

$$J\left[\widetilde{\mathbf{h}}(f)\right] = E\left[\left|\widetilde{\mathcal{E}}(f)\right|^2\right]$$
$$= \phi_{\widetilde{X}}(f) + \widetilde{\mathbf{h}}^H(f)\mathbf{\Phi}_{\mathbf{y}}(f)\widetilde{\mathbf{h}}(f) - \phi_{\widetilde{X}}(f)\widetilde{\mathbf{h}}^H(f)\mathbf{q}_{\mathbf{x},1}(f)$$
$$-\phi_{\widetilde{X}}(f)\mathbf{q}_{\mathbf{x},1}^H(f)\widetilde{\mathbf{h}}(f), \quad (6.51)$$

which can be expressed as

$$J\left[\widetilde{\mathbf{h}}(f)\right] = E\left[\left|\widetilde{\mathcal{E}}_{\mathrm{ds}}(f)\right|^2\right] + E\left[\left|\widetilde{\mathcal{E}}_{\mathrm{rs}}(f)\right|^2\right]$$
$$= J_{\mathrm{ds}}\left[\widetilde{\mathbf{h}}(f)\right] + J_{\mathrm{rs}}\left[\widetilde{\mathbf{h}}(f)\right], \tag{6.52}$$

where

$$J_{\mathrm{ds}}\left[\widetilde{\mathbf{h}}(f)\right] = \phi_{\widetilde{X}}(f)\left|\widetilde{\mathbf{h}}^H(f)\mathbf{q}_{\mathbf{x},1} - 1\right|^2$$
$$= \phi_{\widetilde{X}}(f)\upsilon_{\mathrm{sd}}\left[\widetilde{\mathbf{h}}(f)\right] \tag{6.53}$$

and

$$J_{\mathrm{rs}}\left[\widetilde{\mathbf{h}}(f)\right] = \widetilde{\mathbf{h}}^H(f)\mathbf{\Phi}_{\mathbf{v}}(f)\widetilde{\mathbf{h}}(f)$$
$$= \frac{\mathrm{tr}\left[\mathbf{\Phi}_{\mathbf{v}}(f)\right]}{\xi_{\mathrm{nr}}\left[\widetilde{\mathbf{h}}(f)\right]}. \tag{6.54}$$

We deduce that

$$\frac{J_{\mathrm{ds}}\left[\widetilde{\mathbf{h}}(f)\right]}{J_{\mathrm{rs}}\left[\widetilde{\mathbf{h}}(f)\right]} = \mathrm{iSNR}(f) \times \xi_{\mathrm{nr}}\left[\widetilde{\mathbf{h}}(f)\right] \times \upsilon_{\mathrm{sd}}\left[\widetilde{\mathbf{h}}(f)\right]$$
$$= \mathrm{oSNR}\left[\widetilde{\mathbf{h}}(f)\right] \times \xi_{\mathrm{sr}}\left[\widetilde{\mathbf{h}}(f)\right] \times \upsilon_{\mathrm{sd}}\left[\widetilde{\mathbf{h}}(f)\right]. \tag{6.55}$$

We observe how the narrowband MSEs are related to the narrowband performance measures.

6.4 OPTIMAL FILTERS

In this section, we derive the most important filters that can help mitigate the level of the noise picked up by the microphones. Since we can estimate the whole vector $\mathbf{x}(f)$, the framework proposed here is valid for the monaural and binaural noise reduction problems. In the monaural case, we can choose any component of $\mathbf{x}(f)$ as the desired signal while in the binaural case, any two elements of $\mathbf{x}(f)$ can be considered as the desired signals, which include all the spatial information needed for our binaural hearing system to be able to localize the source signal.

6.4.1 Maximum SNR

Let us rewrite the narrowband output SNR:

$$\mathrm{oSNR}\left[\widetilde{\mathbf{h}}(f)\right] = \frac{\phi_{\widetilde{X}}(f)\widetilde{\mathbf{h}}^H(f)\mathbf{q}_{\mathbf{x},1}(f)\mathbf{q}_{\mathbf{x},1}^H(f)\widetilde{\mathbf{h}}(f)}{\widetilde{\mathbf{h}}^H(f)\mathbf{\Phi}_{\mathbf{v}}(f)\widetilde{\mathbf{h}}(f)}. \tag{6.56}$$

The maximum SNR filter, $\widetilde{\mathbf{h}}_{\max}(f)$, is obtained by maximizing the output SNR as given above. In (6.56), we recognize the generalized Rayleigh quotient [6]. It is well known that this quotient is maximized with the maximum eigenvector of the matrix $\phi_{\widetilde{X}}(f)\mathbf{\Phi}_{\mathbf{v}}^{-1}(f)\mathbf{q}_{\mathbf{x},1}(f)\mathbf{q}_{\mathbf{x},1}^{H}(f)$. Let us denote by $\lambda_{\max}(f)$ the maximum eigenvalue corresponding to this maximum eigenvector. Since the rank of the mentioned matrix is equal to 1, we have

$$\begin{aligned}\lambda_{\max}(f) &= \mathrm{tr}\left[\phi_{\widetilde{X}}(f)\mathbf{\Phi}_{\mathbf{v}}^{-1}(f)\mathbf{q}_{\mathbf{x},1}(f)\mathbf{q}_{\mathbf{x},1}^{H}(f)\right] \\ &= \phi_{\widetilde{X}}(f)\mathbf{q}_{\mathbf{x},1}^{H}(f)\mathbf{\Phi}_{\mathbf{v}}^{-1}(f)\mathbf{q}_{\mathbf{x},1}(f) \\ &= \mathrm{oSNR}_{\max}(f).\end{aligned} \tag{6.57}$$

As a result,

$$\mathrm{oSNR}\left[\widetilde{\mathbf{h}}_{\max}(f)\right] = \lambda_{\max}(f), \tag{6.58}$$

which corresponds to the maximum possible SNR and

$$\begin{aligned}\mathcal{A}\left[\widetilde{\mathbf{h}}_{\max}(f)\right] &= \mathcal{A}_{\max}(f) \\ &= \mathrm{tr}\left[\mathbf{\Phi}_{\mathbf{v}}(f)\right]\mathbf{q}_{\mathbf{x},1}^{H}(f)\mathbf{\Phi}_{\mathbf{v}}^{-1}(f)\mathbf{q}_{\mathbf{x},1}(f).\end{aligned} \tag{6.59}$$

Obviously, we also have

$$\begin{aligned}\widetilde{\mathbf{h}}_{\max}(f) &= \varsigma(f)\mathbf{\Phi}_{\mathbf{v}}^{-1}(f)\mathbf{q}_{\mathbf{x},1}(f) \\ &= \frac{\varsigma(f)}{\sqrt{\mathbf{g}(f)\mathbf{g}^{H}(f)}}\mathbf{\Phi}_{\mathbf{v}}^{-1}(f)\mathbf{g}(f),\end{aligned} \tag{6.60}$$

where $\varsigma(f)$ is an arbitrary frequency-dependent complex number different from zero. While this factor has no effect on the narrowband output SNR, it has on the broadband output SNR and on the speech distortion. In fact, all the filters (except for the LCMV) derived in the rest of this section are equivalent up to this complex factor. These filters also try to find the respective complex factors at each frequency depending on what we optimize. It is important to understand that while the maximum SNR filter maximizes the narrowband output SNR, it certainly does not maximize the broadband output SNR whose value depends quite a lot on the $\varsigma(f)$'s.

Let us denote by $\mathcal{A}_{\max}^{(m)}(f)$ the maximum narrowband array gain of a microphone array with m sensors. By virtue of the inclusion principle [6] for the matrix $\phi_{\widetilde{X}}(f)\mathbf{\Phi}_{\mathbf{v}}^{-1}(f)\mathbf{q}_{\mathbf{x},1}(f)\mathbf{q}_{\mathbf{x},1}^{H}(f)$, we have

$$\mathcal{A}_{\max}^{(M)}(f) \geq \mathcal{A}_{\max}^{(M-1)}(f) \geq \cdots \geq \mathcal{A}_{\max}^{(2)}(f) \geq \mathcal{A}_{\max}^{(1)}(f) = 1. \tag{6.61}$$

This shows that by increasing the number of microphones, we necessarily increase the narrowband array gain. If there is one microphone only, the narrowband array gain cannot be improved as expected [7].

Using (6.22), we deduce the optimal, in the maximum SNR sense, equivalent filtering matrix for the estimation of the whole vector $\mathbf{x}(f)$:

$$\begin{aligned}\mathbf{H}_{\max}(f) &= \varsigma^*(f)\mathbf{q}_{\mathbf{x},1}(f)\mathbf{q}_{\mathbf{x},1}^H(f)\mathbf{\Phi}_{\mathbf{v}}^{-1}(f) \\ &= \frac{\varsigma^*(f)}{\phi_{\widetilde{X}}(f)}\mathbf{\Phi}_{\mathbf{x}}(f)\mathbf{\Phi}_{\mathbf{v}}^{-1}(f) \\ &= \frac{\varsigma^*(f)}{\phi_{\widetilde{X}}(f)}\left[\mathbf{\Phi}_{\mathbf{y}}(f)\mathbf{\Phi}_{\mathbf{v}}^{-1}(f) - \mathbf{I}_M\right]. \end{aligned} \tag{6.62}$$

We deduce that any element, $X_m(f)$, of the vector $\mathbf{x}(f)$ can be estimated by the filter:

$$\mathbf{h}_{\max,m}(f) = \frac{\varsigma(f)}{\phi_{\widetilde{X}}(f)}\left[\mathbf{\Phi}_{\mathbf{v}}^{-1}(f)\mathbf{\Phi}_{\mathbf{y}}(f) - \mathbf{I}_M\right]\mathbf{i}_m. \tag{6.63}$$

6.4.2 Wiener

By minimizing the narrowband MSE, $J\left[\widetilde{\mathbf{h}}(f)\right]$, with respect to $\widetilde{\mathbf{h}}(f)$, we easily find the Wiener filter:

$$\widetilde{\mathbf{h}}_{\mathrm{W}}(f) = \phi_{\widetilde{X}}(f)\mathbf{\Phi}_{\mathbf{y}}^{-1}(f)\mathbf{q}_{\mathbf{x},1}(f). \tag{6.64}$$

Let

$$\mathbf{\Gamma}_{\mathbf{y}}(f) = M\frac{\mathbf{\Phi}_{\mathbf{y}}(f)}{\mathrm{tr}\left[\mathbf{\Phi}_{\mathbf{y}}(f)\right]} \tag{6.65}$$

be the pseudo-coherence matrix of the spatial observations, we can rewrite (6.64) as

$$\begin{aligned}\widetilde{\mathbf{h}}_{\mathrm{W}}(f) &= M\frac{\mathrm{iSNR}(f)}{1+\mathrm{iSNR}(f)}\mathbf{\Gamma}_{\mathbf{y}}^{-1}(f)\mathbf{q}_{\mathbf{x},1}(f) \\ &= MH_{\mathrm{W}}(f)\mathbf{\Gamma}_{\mathbf{y}}^{-1}(f)\mathbf{q}_{\mathbf{x},1}(f), \end{aligned} \tag{6.66}$$

where

$$H_{\mathrm{W}}(f) = \frac{\mathrm{iSNR}(f)}{1+\mathrm{iSNR}(f)} \tag{6.67}$$

is the (single-channel) Wiener gain and $\mathbf{\Gamma}_{\mathbf{y}}^{-1}(f)\mathbf{q}_{\mathbf{x},1}(f)$ is the spatial information vector. If we do not want to rely on the statistics of the noise to estimate the Wiener filter, we can approximate $\mathbf{q}_{\mathbf{x},1}(f)$ with the steering vector and estimate $\text{iSNR}(f)$ with the decision-directed approach [8].

We can write the general form of the Wiener filter in another way that will make it easier to compare to other optimal filters. Indeed, determining the inverse of $\mathbf{\Phi}_{\mathbf{y}}(f)$ from (6.15) with the Woodbury's identity, we get

$$\mathbf{\Phi}_{\mathbf{y}}^{-1}(f) = \mathbf{\Phi}_{\mathbf{v}}^{-1}(f) - \frac{\mathbf{\Phi}_{\mathbf{v}}^{-1}(f)\mathbf{q}_{\mathbf{x},1}(f)\mathbf{q}_{\mathbf{x},1}^{H}(f)\mathbf{\Phi}_{\mathbf{v}}^{-1}(f)}{\phi_{\widetilde{X}}^{-1}(f) + \mathbf{q}_{\mathbf{x},1}^{H}(f)\mathbf{\Phi}_{\mathbf{v}}^{-1}(f)\mathbf{q}_{\mathbf{x},1}(f)}. \tag{6.68}$$

Substituting (6.68) into (6.64) gives

$$\begin{aligned}\widetilde{\mathbf{h}}_{\mathrm{W}}(f) &= \frac{\phi_{\widetilde{X}}(f)\mathbf{\Phi}_{\mathbf{v}}^{-1}(f)\mathbf{q}_{\mathbf{x},1}(f)}{1 + \phi_{\widetilde{X}}(f)\mathbf{q}_{\mathbf{x},1}^{H}(f)\mathbf{\Phi}_{\mathbf{v}}^{-1}(f)\mathbf{q}_{\mathbf{x},1}(f)} \\ &= \frac{\phi_{\widetilde{X}}(f)\mathbf{\Phi}_{\mathbf{v}}^{-1}(f)\mathbf{q}_{\mathbf{x},1}(f)}{1 + \lambda_{\max}(f)}.\end{aligned} \tag{6.69}$$

It is interesting to see that the two filters $\widetilde{\mathbf{h}}_{\mathrm{W}}(f)$ and $\widetilde{\mathbf{h}}_{\max}(f)$ differ only by a real-valued factor. Indeed, taking

$$\varsigma(f) = \frac{\phi_{\widetilde{X}}(f)}{1 + \lambda_{\max}(f)} \tag{6.70}$$

in (6.60) (maximum SNR filter), we find (6.69) (Wiener filter).

From (6.69), we deduce that the narrowband output SNR is

$$\begin{aligned}\text{oSNR}\left[\widetilde{\mathbf{h}}_{\mathrm{W}}(f)\right] &= \lambda_{\max}(f) \\ &= \text{tr}\left[\mathbf{\Phi}_{\mathbf{v}}^{-1}(f)\mathbf{\Phi}_{\mathbf{y}}(f)\right] - M\end{aligned} \tag{6.71}$$

and, obviously,

$$\text{oSNR}\left[\widetilde{\mathbf{h}}_{\mathrm{W}}(f)\right] \geq \text{iSNR}(f), \tag{6.72}$$

since the Wiener filter maximizes the narrowband output SNR.

The speech distortion indices are

$$\upsilon_{\text{sd}}\left[\widetilde{\mathbf{h}}_{\mathrm{W}}(f)\right] = \frac{1}{\left[1 + \lambda_{\max}(f)\right]^2}, \tag{6.73}$$

$$\upsilon_{\mathrm{sd}}\left(\widetilde{\mathbf{h}}_{\mathrm{W}}\right) = \frac{\int_f \phi_{\widetilde{X}}(f)\left[1+\lambda_{\max}(f)\right]^{-2} df}{\int_f \phi_{\widetilde{X}}(f) df}. \quad (6.74)$$

The higher the value of $\lambda_{\max}(f)$ (and/or the number of microphones), the less the desired signal is distorted.

It is also easy to find the noise reduction factors:

$$\xi_{\mathrm{nr}}\left[\widetilde{\mathbf{h}}_{\mathrm{W}}(f)\right] = \frac{\left[1+\lambda_{\max}(f)\right]^2}{\mathrm{iSNR}(f)\lambda_{\max}(f)}, \quad (6.75)$$

$$\xi_{\mathrm{nr}}\left(\widetilde{\mathbf{h}}_{\mathrm{W}}\right) = \frac{\int_f \phi_{\widetilde{X}}(f)\mathrm{iSNR}^{-1}(f) df}{\int_f \phi_{\widetilde{X}}(f)\lambda_{\max}(f)\left[1+\lambda_{\max}(f)\right]^{-2} df}, \quad (6.76)$$

and the speech reduction factors:

$$\xi_{\mathrm{sr}}\left[\widetilde{\mathbf{h}}_{\mathrm{W}}(f)\right] = \frac{\left[1+\lambda_{\max}(f)\right]^2}{\lambda_{\max}^2(f)}, \quad (6.77)$$

$$\xi_{\mathrm{sr}}\left(\widetilde{\mathbf{h}}_{\mathrm{W}}\right) = \frac{\int_f \phi_{\widetilde{X}}(f) df}{\int_f \phi_{\widetilde{X}}(f)\lambda_{\max}^2(f)\left[1+\lambda_{\max}(f)\right]^{-2} df}. \quad (6.78)$$

The broadband output SNR of the Wiener filter is

$$\mathrm{oSNR}\left(\widetilde{\mathbf{h}}_{\mathrm{W}}\right) = \frac{\displaystyle\int_f \phi_{\widetilde{X}}(f)\frac{\lambda_{\max}^2(f)}{\left[1+\lambda_{\max}(f)\right]^2} df}{\displaystyle\int_f \phi_{\widetilde{X}}(f)\frac{\lambda_{\max}(f)}{\left[1+\lambda_{\max}(f)\right]^2} df}. \quad (6.79)$$

Property 6.1. *With the multichannel frequency-domain Wiener filter given in (6.64), the broadband output SNR is always greater than or equal to the broadband input SNR, i.e.,* $\mathrm{oSNR}\left(\widetilde{\mathbf{h}}_{\mathrm{W}}\right) \geq \mathrm{iSNR}$.

From (6.22) and (6.64), we deduce the Wiener filtering matrix for the estimation of the vector $\mathbf{x}(f)$:

$$\begin{aligned}\mathbf{H}_{\mathrm{W}}(f) &= \phi_{\widetilde{X}}(f)\mathbf{q}_{\mathbf{x},1}(f)\mathbf{q}_{\mathbf{x},1}^H(f)\mathbf{\Phi}_{\mathbf{y}}^{-1}(f) \\ &= \mathbf{\Phi}_{\mathbf{x}}(f)\mathbf{\Phi}_{\mathbf{y}}^{-1}(f) \\ &= \mathbf{I}_M - \mathbf{\Phi}_{\mathbf{v}}(f)\mathbf{\Phi}_{\mathbf{y}}^{-1}(f)\end{aligned} \quad (6.80)$$

or, from (6.69),

$$\mathbf{H}_{\text{W}}(f) = \frac{\mathbf{\Phi}_{\mathbf{x}}(f)\mathbf{\Phi}_{\mathbf{v}}^{-1}(f)}{1 + \text{tr}\left[\mathbf{\Phi}_{\mathbf{v}}^{-1}(f)\mathbf{\Phi}_{\mathbf{x}}(f)\right]} = \frac{\mathbf{\Phi}_{\mathbf{y}}(f)\mathbf{\Phi}_{\mathbf{v}}^{-1}(f) - \mathbf{I}_M}{1 - M + \text{tr}\left[\mathbf{\Phi}_{\mathbf{v}}^{-1}(f)\mathbf{\Phi}_{\mathbf{y}}(f)\right]}. \tag{6.81}$$

As a result, $X_m(f)$ can be estimated with the Wiener filter:

$$\mathbf{h}_{\text{W},m}(f) = \left[\mathbf{I}_M - \mathbf{\Phi}_{\mathbf{y}}^{-1}(f)\mathbf{\Phi}_{\mathbf{v}}(f)\right]\mathbf{i}_m = \frac{\mathbf{\Phi}_{\mathbf{v}}^{-1}(f)\mathbf{\Phi}_{\mathbf{y}}(f) - \mathbf{I}_M}{1 - M + \text{tr}\left[\mathbf{\Phi}_{\mathbf{v}}^{-1}(f)\mathbf{\Phi}_{\mathbf{y}}(f)\right]}\mathbf{i}_m. \tag{6.82}$$

We can express $\mathbf{h}_{\text{W},m}(f)$ as a function of the narrowband input SNR and the pseudo-coherence matrices, i.e.,

$$\mathbf{h}_{\text{W},m}(f) = \frac{\left[1 + \text{iSNR}(f)\right]\mathbf{\Gamma}_{\mathbf{v}}^{-1}(f)\mathbf{\Gamma}_{\mathbf{y}}(f) - \mathbf{I}_M}{1 - M + \left[1 + \text{iSNR}(f)\right]\text{tr}\left[\mathbf{\Gamma}_{\mathbf{v}}^{-1}(f)\mathbf{\Gamma}_{\mathbf{y}}(f)\right]}\mathbf{i}_m, \tag{6.83}$$

where

$$\mathbf{\Gamma}_{\mathbf{v}}(f) = M\frac{\mathbf{\Phi}_{\mathbf{v}}(f)}{\text{tr}\left[\mathbf{\Phi}_{\mathbf{v}}(f)\right]}. \tag{6.84}$$

If we know that we are in the presence of the spherically isotropic noise, the Wiener filter simplifies to

$$\mathbf{h}_{\text{W},m}(f) = \frac{\left[1 + \text{iSNR}(f)\right]\mathbf{\Gamma}_{\text{si}}^{-1}(f)\mathbf{\Gamma}_{\mathbf{y}}(f) - \mathbf{I}_M}{1 - M + \left[1 + \text{iSNR}(f)\right]\text{tr}\left[\mathbf{\Gamma}_{\text{si}}^{-1}(f)\mathbf{\Gamma}_{\mathbf{y}}(f)\right]}\mathbf{i}_m, \tag{6.85}$$

which makes it very practical since the coherence matrix, $\mathbf{\Gamma}_{\text{si}}(f)$, of the spherically isotropic noise is known, while $\text{iSNR}(f)$ and $\mathbf{\Gamma}_{\mathbf{y}}(f)$ are easy to estimate.

6.4.3 MVDR

The well-known MVDR filter proposed by Capon [9, 10] is derived by minimizing the narrowband MSE of the residual noise with the distortionless constraint, i.e.,

$$\min_{\widetilde{\mathbf{h}}(f)} \widetilde{\mathbf{h}}^H(f)\mathbf{\Phi}_{\mathbf{v}}(f)\widetilde{\mathbf{h}}(f) \quad \text{subject to} \quad \widetilde{\mathbf{h}}^H(f)\mathbf{q}_{\mathbf{x},1}(f) = 1, \tag{6.86}$$

for which the solution is

$$\widetilde{\mathbf{h}}_{\text{MVDR}}(f) = \frac{\mathbf{\Phi}_{\mathbf{v}}^{-1}(f)\mathbf{q}_{\mathbf{x},1}(f)}{\mathbf{q}_{\mathbf{x},1}^{H}(f)(f)\mathbf{\Phi}_{\mathbf{v}}^{-1}(f)\mathbf{q}_{\mathbf{x},1}(f)} = \frac{\phi_{\widetilde{X}}(f)\mathbf{\Phi}_{\mathbf{v}}^{-1}(f)\mathbf{q}_{\mathbf{x},1}(f)}{\lambda_{\max}(f)}. \tag{6.87}$$

Taking

$$\varsigma(f) = \frac{\phi_{\widetilde{X}}(f)}{\lambda_{\max}(f)} \tag{6.88}$$

in (6.60) (maximum SNR filter), we find (6.87) (MVDR filter), showing how the maximum SNR and MVDR filters are equivalent up to a real-valued factor. Alternatively, we can also write the MVDR as

$$\widetilde{\mathbf{h}}_{\text{MVDR}}(f) = \frac{\mathbf{\Phi}_{\mathbf{y}}^{-1}(f)\mathbf{q}_{\mathbf{x},1}(f)}{\mathbf{q}_{\mathbf{x},1}^{H}(f)(f)\mathbf{\Phi}_{\mathbf{y}}^{-1}(f)\mathbf{q}_{\mathbf{x},1}(f)}. \tag{6.89}$$

The Wiener and MVDR filters are simply related as follows:

$$\widetilde{\mathbf{h}}_{\text{W}}(f) = C_{\text{W}}(f)\widetilde{\mathbf{h}}_{\text{MVDR}}(f), \tag{6.90}$$

where

$$C_{\text{W}}(f) = \widetilde{\mathbf{h}}_{\text{W}}^{H}(f)\mathbf{q}_{\mathbf{x},1}(f) = \frac{\lambda_{\max}(f)}{1+\lambda_{\max}(f)} \tag{6.91}$$

can be seen as a single-channel frequency-domain Wiener gain. In fact, any filter of the form:

$$\widetilde{\mathbf{h}}(f) = C(f)\widetilde{\mathbf{h}}_{\text{MVDR}}(f), \tag{6.92}$$

where $C(f)$ is a real number with $0 < C(f) < 1$, removes more noise than the MVDR filter at the price of some desired signal distortion, which is

$$\xi_{\text{sr}}\left[\widetilde{\mathbf{h}}(f)\right] = \frac{1}{C^2(f)} \tag{6.93}$$

or

$$\upsilon_{\text{sd}}\left[\widetilde{\mathbf{h}}(f)\right] = \left[C(f) - 1\right]^2. \tag{6.94}$$

It can be verified that we always have

$$\text{oSNR}\left[\widetilde{\mathbf{h}}_{\text{MVDR}}(f)\right] = \text{oSNR}\left[\widetilde{\mathbf{h}}_{\text{W}}(f)\right], \tag{6.95}$$

$$\upsilon_{\text{sd}}\left[\widetilde{\mathbf{h}}_{\text{MVDR}}(f)\right] = 0, \tag{6.96}$$

$$\xi_{\text{sr}}\left[\widetilde{\mathbf{h}}_{\text{MVDR}}(f)\right] = 1, \tag{6.97}$$

and

$$\xi_{\text{nr}}\left[\widetilde{\mathbf{h}}_{\text{MVDR}}(f)\right] \leq \xi_{\text{nr}}\left[\widetilde{\mathbf{h}}_{\text{W}}(f)\right], \tag{6.98}$$

$$\xi_{\text{nr}}\left(\widetilde{\mathbf{h}}_{\text{MVDR}}\right) \leq \xi_{\text{nr}}\left(\widetilde{\mathbf{h}}_{\text{W}}\right). \tag{6.99}$$

The MVDR filter rejects the maximum level of noise allowable without distorting the desired signal at each frequency.

While the narrowband output SNRs of the Wiener and MVDR filters are strictly equal, their broadband output SNRs are not. The broadband output SNR of the MVDR filter is

$$\text{oSNR}\left(\widetilde{\mathbf{h}}_{\text{MVDR}}\right) = \frac{\int_f \phi_{\widetilde{X}}(f)df}{\int_f \phi_{\widetilde{X}}(f)\lambda_{\max}^{-1}(f)df} \tag{6.100}$$

and

$$\text{oSNR}\left(\widetilde{\mathbf{h}}_{\text{MVDR}}\right) \leq \text{oSNR}\left(\widetilde{\mathbf{h}}_{\text{W}}\right). \tag{6.101}$$

Property 6.2. *With the multichannel frequency-domain MVDR filter given in (6.87), the broadband output SNR is always greater than or equal to the broadband input SNR, i.e.,* $\text{oSNR}\left(\widetilde{\mathbf{h}}_{\text{MVDR}}\right) \geq \text{iSNR}$.

It is easy to observe that the MVDR filtering matrix for the estimation of the vector $\mathbf{x}(f)$ is

$$\begin{aligned}\mathbf{H}_{\text{MVDR}}(f) &= \frac{\mathbf{q}_{\mathbf{x},1}(f)\mathbf{q}_{\mathbf{x},1}^H(f)\mathbf{\Phi}_{\mathbf{v}}^{-1}(f)}{\mathbf{q}_{\mathbf{x},1}^H(f)(f)\mathbf{\Phi}_{\mathbf{v}}^{-1}(f)\mathbf{q}_{\mathbf{x},1}(f)} \\ &= \frac{\mathbf{\Phi}_{\mathbf{x}}(f)\mathbf{\Phi}_{\mathbf{v}}^{-1}(f)}{\text{tr}\left[\mathbf{\Phi}_{\mathbf{v}}^{-1}(f)\mathbf{\Phi}_{\mathbf{x}}(f)\right]}\end{aligned}$$

$$= \frac{\mathbf{\Phi}_{\mathbf{y}}(f)\mathbf{\Phi}_{\mathbf{v}}^{-1}(f) - \mathbf{I}_M}{\text{tr}\left[\mathbf{\Phi}_{\mathbf{v}}^{-1}(f)\mathbf{\Phi}_{\mathbf{y}}(f)\right] - M} \tag{6.102}$$

or, from (6.89),

$$\mathbf{H}_{\text{MVDR}}(f) = \frac{\mathbf{I}_M - \mathbf{\Phi}_{\mathbf{v}}(f)\mathbf{\Phi}_{\mathbf{y}}^{-1}(f)}{M - \text{tr}\left[\mathbf{\Phi}_{\mathbf{y}}^{-1}(f)\mathbf{\Phi}_{\mathbf{v}}(f)\right]}$$

$$= \frac{\mathbf{H}_{\text{W}}(f)}{M - \text{tr}\left[\mathbf{\Phi}_{\mathbf{y}}^{-1}(f)\mathbf{\Phi}_{\mathbf{v}}(f)\right]}. \tag{6.103}$$

As a result, $X_m(f)$ can be estimated with the MVDR filter:

$$\mathbf{h}_{\text{MVDR},m}(f) = \frac{\mathbf{I}_M - \mathbf{\Phi}_{\mathbf{v}}(f)\mathbf{\Phi}_{\mathbf{y}}^{-1}(f)}{M - \text{tr}\left[\mathbf{\Phi}_{\mathbf{y}}^{-1}(f)\mathbf{\Phi}_{\mathbf{v}}(f)\right]}\mathbf{i}_m$$

$$= \frac{\mathbf{\Phi}_{\mathbf{v}}^{-1}(f)\mathbf{\Phi}_{\mathbf{y}}(f) - \mathbf{I}_M}{\text{tr}\left[\mathbf{\Phi}_{\mathbf{v}}^{-1}(f)\mathbf{\Phi}_{\mathbf{y}}(f)\right] - M}\mathbf{i}_m. \tag{6.104}$$

We can express $\mathbf{h}_{\text{MVDR},m}(f)$ as a function of the narrowband input SNR and the pseudo-coherence matrices, i.e.,

$$\mathbf{h}_{\text{W},m}(f) = \frac{\left[1 + \text{iSNR}(f)\right]\mathbf{\Gamma}_{\mathbf{v}}^{-1}(f)\mathbf{\Gamma}_{\mathbf{y}}(f) - \mathbf{I}_M}{\left[1 + \text{iSNR}(f)\right]\text{tr}\left[\mathbf{\Gamma}_{\mathbf{v}}^{-1}(f)\mathbf{\Gamma}_{\mathbf{y}}(f)\right] - M}\mathbf{i}_m. \tag{6.105}$$

If we know that we are in the presence of the spherically isotropic noise, the MVDR filter simplifies to

$$\mathbf{h}_{\text{MVDR},m}(f) = \frac{\left[1 + \text{iSNR}(f)\right]\mathbf{\Gamma}_{\text{si}}^{-1}(f)\mathbf{\Gamma}_{\mathbf{y}}(f) - \mathbf{I}_M}{\left[1 + \text{iSNR}(f)\right]\text{tr}\left[\mathbf{\Gamma}_{\text{si}}^{-1}(f)\mathbf{\Gamma}_{\mathbf{y}}(f)\right] - M}\mathbf{i}_m. \tag{6.106}$$

6.4.4 Tradeoff

The tradeoff filter is derived by minimizing the narrowband MSE of the speech distortion with the constraint that the narrowband noise reduction factor is equal to a positive value that is greater than 1, i.e.,

$$\min_{\widetilde{\mathbf{h}}(f)} J_{\text{ds}}\left[\widetilde{\mathbf{h}}(f)\right] \quad \text{subject to} \quad J_{\text{rs}}\left[\widetilde{\mathbf{h}}(f)\right] = \beta\text{tr}\left[\mathbf{\Phi}_{\mathbf{v}}(f)\right], \tag{6.107}$$

where $0 < \beta < 1$ to insure that we get some noise reduction. By using a Lagrange multiplier, $\mu > 0$, to adjoin the constraint to the cost function, we easily deduce the tradeoff filter:

$$\begin{aligned}\widetilde{\mathbf{h}}_{\mathrm{T},\mu}(f) &= \phi_{\widetilde{X}}(f)\left[\mathbf{\Phi}_{\mathbf{x}}(f) + \mu\mathbf{\Phi}_{\mathbf{v}}(f)\right]^{-1}\mathbf{q}_{\mathbf{x},1}(f) \\ &= \frac{\phi_{\widetilde{X}}(f)\mathbf{\Phi}_{\mathbf{v}}^{-1}(f)\mathbf{q}_{\mathbf{x},1}(f)}{\mu + \phi_{\widetilde{X}}(f)\mathbf{q}_{\mathbf{x},1}^{H}(f)\mathbf{\Phi}_{\mathbf{v}}^{-1}(f)\mathbf{q}_{\mathbf{x},1}(f)},\end{aligned} \tag{6.108}$$

where we have assumed that the matrix $\mathbf{\Phi}_{\mathbf{x}}(f)+\mu\mathbf{\Phi}_{\mathbf{v}}(f)$ is invertible and the Lagrange multiplier, μ, satisfies

$$J_{\mathrm{rs}}\left[\widetilde{\mathbf{h}}_{\mathrm{T},\mu}(f)\right] = \beta\mathrm{tr}\left[\mathbf{\Phi}_{\mathbf{v}}(f)\right]. \tag{6.109}$$

However, in practice, it is not easy to determine the optimal μ. Therefore, when this parameter is chosen in a heuristic way, we can see that for

- $\mu = 1$, $\widetilde{\mathbf{h}}_{\mathrm{T},1}(f) = \widetilde{\mathbf{h}}_{\mathrm{W}}(f)$, which is the Wiener filter;
- $\mu = 0$, $\widetilde{\mathbf{h}}_{\mathrm{T},0}(f) = \widetilde{\mathbf{h}}_{\mathrm{MVDR}}(f)$, which is the MVDR filter;
- $\mu > 1$, results in a filter with low residual noise at the expense of high speech distortion (as compared to Wiener);
- $\mu < 1$, results in a filter with high residual noise and low speech distortion (as compared to Wiener).

Note that the MVDR cannot be derived from the first line of (6.108) since by taking $\mu = 0$, we have to invert a matrix that is not full rank.

It can be observed that the tradeoff, Wiener, and maximum SNR filters are equivalent up to a real-valued number. As a result, the narrowband output SNR of the tradeoff filter is independent of μ and is identical to the narrowband output SNR of the maximum SNR filter, i.e.,

$$\mathrm{oSNR}\left[\widetilde{\mathbf{h}}_{\mathrm{T},\mu}(f)\right] = \mathrm{oSNR}\left[\widetilde{\mathbf{h}}_{\mathrm{max}}(f)\right], \quad \forall\mu \geq 0. \tag{6.110}$$

We have

$$\upsilon_{\mathrm{sd}}\left[\widetilde{\mathbf{h}}_{\mathrm{T},\mu}(f)\right] = \left[\frac{\mu}{\mu + \lambda_{\mathrm{max}}(f)}\right]^2, \tag{6.111}$$

$$\xi_{\mathrm{sr}}\left[\widetilde{\mathbf{h}}_{\mathrm{T},\mu}(f)\right] = \left[1 + \frac{\mu}{\lambda_{\mathrm{max}}(f)}\right]^2, \tag{6.112}$$

$$\xi_{\mathrm{nr}}\left[\widetilde{\mathbf{h}}_{\mathrm{T},\mu}(f)\right] = \frac{\left[\mu + \lambda_{\mathrm{max}}(f)\right]^2}{\mathrm{iSNR}(f)\lambda_{\mathrm{max}}(f)}. \tag{6.113}$$

The tradeoff filter is interesting from several perspectives since it encompasses both the Wiener and MVDR filters. It is then useful to study the broadband output SNR and the broadband speech distortion index of the tradeoff filter. Next, we give some important results.

It can be verified that the broadband output SNR of the tradeoff filter is

$$\mathrm{oSNR}\left(\widetilde{\mathbf{h}}_{\mathrm{T},\mu}\right) = \frac{\displaystyle\int_f \phi_{\widetilde{X}}(f)\frac{\lambda_{\max}^2(f)}{\left[\mu + \lambda_{\max}(f)\right]^2}df}{\displaystyle\int_f \phi_{\widetilde{X}}(f)\frac{\lambda_{\max}(f)}{\left[\mu + \lambda_{\max}(f)\right]^2}df}. \tag{6.114}$$

Property 6.3. *The broadband output SNR of the tradeoff filter is an increasing function of the parameter* μ [11].

An important consequence of the previous property is that in the class of the tradeoff filters, the MVDR filter gives the smallest broadband output SNR [11].

While the broadband output SNR is upper bounded, it is easy to see that the broadband noise reduction factor and broadband speech reduction factor are not. So when μ goes to infinity, so are $\xi_{\mathrm{nr}}\left(\widetilde{\mathbf{h}}_{\mathrm{T},\mu}\right)$ and $\xi_{\mathrm{sr}}\left(\widetilde{\mathbf{h}}_{\mathrm{T},\mu}\right)$.

The broadband speech distortion index is

$$\upsilon_{\mathrm{sd}}\left(\widetilde{\mathbf{h}}_{\mathrm{T},\mu}\right) = \frac{\int_f \phi_{\widetilde{X}}(f)\frac{\mu^2}{\left[\mu+\lambda_{\max}(f)\right]^2}df}{\int_f \phi_{\widetilde{X}}(f)df}. \tag{6.115}$$

Property 6.4. *The broadband speech distortion index of the tradeoff filter is an increasing function of the parameter* μ.

It is clear that

$$0 \le \upsilon_{\mathrm{sd}}\left(\widetilde{\mathbf{h}}_{\mathrm{T},\mu}\right) \le 1, \forall\mu \ge 0. \tag{6.116}$$

Therefore, as μ increases, the broadband output SNR increases at the price of more distortion to the desired signal.

Property 6.5. *With the multichannel frequency-domain tradeoff filter given in (6.108), the broadband output SNR is always greater than or equal to the broadband input SNR, i.e.,* $\mathrm{oSNR}\left(\widetilde{\mathbf{h}}_{\mathrm{T},\mu}\right) \ge \mathrm{iSNR},\ \forall\mu \ge 0$ [11].

From the previous results, we deduce that, for $\mu \geq 1$,

$$1 \leq \mathcal{A}\left(\widetilde{\mathbf{h}}_{\text{MVDR}}\right) \leq \mathcal{A}\left(\widetilde{\mathbf{h}}_{\text{W}}\right) \leq \mathcal{A}\left(\widetilde{\mathbf{h}}_{\text{T},\mu}\right), \tag{6.117}$$

$$0 = \upsilon_{\text{sd}}\left(\widetilde{\mathbf{h}}_{\text{MVDR}}\right) \leq \upsilon_{\text{sd}}\left(\widetilde{\mathbf{h}}_{\text{W}}\right) \leq \upsilon_{\text{sd}}\left(\widetilde{\mathbf{h}}_{\text{T},\mu}\right), \tag{6.118}$$

and, for $0 \leq \mu \leq 1$,

$$1 \leq \mathcal{A}\left(\widetilde{\mathbf{h}}_{\text{MVDR}}\right) \leq \mathcal{A}\left(\widetilde{\mathbf{h}}_{\text{T},\mu}\right) \leq \mathcal{A}\left(\widetilde{\mathbf{h}}_{\text{W}}\right), \tag{6.119}$$

$$0 = \upsilon_{\text{sd}}\left(\widetilde{\mathbf{h}}_{\text{MVDR}}\right) \leq \upsilon_{\text{sd}}\left(\widetilde{\mathbf{h}}_{\text{T},\mu}\right) \leq \upsilon_{\text{sd}}\left(\widetilde{\mathbf{h}}_{\text{W}}\right). \tag{6.120}$$

It can be verified that the equivalent tradeoff filtering matrix for the estimation of $\mathbf{x}(f)$ is

$$\mathbf{H}_{\text{T},\mu}(f) = \frac{\mathbf{\Phi}_{\mathbf{y}}(f)\mathbf{\Phi}_{\mathbf{v}}^{-1}(f) - \mathbf{I}_M}{\mu - M + \text{tr}\left[\mathbf{\Phi}_{\mathbf{v}}^{-1}(f)\mathbf{\Phi}_{\mathbf{y}}(f)\right]}, \tag{6.121}$$

so that the tradeoff filter to estimate $X_m(f)$ is

$$\mathbf{h}_{\text{T},\mu,m}(f) = \frac{\mathbf{\Phi}_{\mathbf{v}}^{-1}(f)\mathbf{\Phi}_{\mathbf{y}}(f) - \mathbf{I}_M}{\mu - M + \text{tr}\left[\mathbf{\Phi}_{\mathbf{v}}^{-1}(f)\mathbf{\Phi}_{\mathbf{y}}(f)\right]}\mathbf{i}_m. \tag{6.122}$$

6.4.5 LCMV

It is possible to derive an LCMV filter [12, 13], which can handle more than one linear constraint, by exploiting the decomposition of the noise spatial correlation matrix. Let

$$\mathbf{\Phi}_{\mathbf{v}}(f) = \mathbf{Q}_{\mathbf{v}}(f)\mathbf{\Lambda}_{\mathbf{v}}(f)\mathbf{Q}_{\mathbf{v}}^H(f), \tag{6.123}$$

where the unitary and diagonal matrices $\mathbf{Q}_{\mathbf{v}}(f)$ and $\mathbf{\Lambda}_{\mathbf{v}}(f)$ are defined similarly to $\mathbf{Q}_{\mathbf{x}}(f)$ and $\mathbf{\Lambda}_{\mathbf{x}}(f)$, respectively. We assume that the (positive) eigenvalues of $\mathbf{\Phi}_{\mathbf{v}}(f)$ have the following structure: $\lambda_{\mathbf{v},1}(f) \geq \lambda_{\mathbf{v},2}(f) \geq \cdots \geq \lambda_{\mathbf{v},Q}(f) > \phi_{\text{swn}}(f)$ and $\lambda_{\mathbf{v},Q+1}(f) = \lambda_{\mathbf{v},Q+2}(f) = \cdots = \lambda_{\mathbf{v},M}(f) = \phi_{\text{swn}}(f)$, where $Q+1 \leq M$. In this case, we can express the unitary matrix as

$$\mathbf{Q}_{\mathbf{v}}(f) = \left[\mathbf{T}_{\mathbf{v}}(f)\ \mathbf{\Upsilon}_{\mathbf{v}}(f)\right], \tag{6.124}$$

where the $M \times Q$ matrix $\mathbf{T}_{\mathbf{v}}(f)$ contains the eigenvectors corresponding to the first Q eigenvalues of $\mathbf{\Phi}_{\mathbf{v}}(f)$ and the $M \times (M - Q)$ matrix $\mathbf{\Upsilon}_{\mathbf{v}}(f)$ contains the eigenvectors corresponding to the last $M - Q$ eigenvalues of $\mathbf{\Phi}_{\mathbf{v}}(f)$. As a result, the noise signal vector can be decomposed as

$$\mathbf{v}(f) = \mathbf{v}_{\text{c}}(f) + \mathbf{v}_{\text{i}}(f), \tag{6.125}$$

where

$$\mathbf{v}_{\mathrm{c}}(f) = \mathbf{T}_{\mathbf{v}}(f)\mathbf{T}_{\mathbf{v}}^{H}(f)\mathbf{v}(f) \tag{6.126}$$

corresponds to the coherent noise,

$$\mathbf{v}_{\mathrm{i}}(f) = \boldsymbol{\Upsilon}_{\mathbf{v}}(f)\boldsymbol{\Upsilon}_{\mathbf{v}}^{H}(f)\mathbf{v}(f) \tag{6.127}$$

corresponds to the incoherent noise, and $E\left[\mathbf{v}_{\mathrm{c}}(f)\mathbf{v}_{\mathrm{i}}^{H}(f)\right] = \mathbf{0}_{M\times M}$.

The LCMV filter that we propose in this subsection consists of estimating $\widetilde{X}(f)$ without any distortion, partially removing the coherent noise, and attenuating the incoherent noise as much as possible. It follows that the constraints are

$$\widetilde{\mathbf{h}}^{H}(f)\mathbf{C}_{\widetilde{X}\mathbf{v}}(f) = \left[1 \; \boldsymbol{\alpha}^{T}\right], \tag{6.128}$$

where

$$\mathbf{C}_{\widetilde{X}\mathbf{v}}(f) = \left[\mathbf{q}_{\mathbf{x},1}(f) \; \mathbf{T}_{\mathbf{v}}(f)\right] \tag{6.129}$$

is the constraint matrix of size $M \times (Q+1)$ and

$$\boldsymbol{\alpha} = \left[\alpha_1 \; \alpha_2 \; \cdots \; \alpha_Q\right]^{T} \tag{6.130}$$

is the attenuation vector of length Q with $0 \leq \alpha_q < 1, \; q = 1, 2, \ldots, Q$. The optimization problem is now

$$\min_{\widetilde{\mathbf{h}}(f)} \widetilde{\mathbf{h}}^{H}(f)\boldsymbol{\Phi}_{\mathbf{y}}(f)\widetilde{\mathbf{h}}(f) \quad \text{subject to} \quad \widetilde{\mathbf{h}}^{H}(f)\mathbf{C}_{\widetilde{X}\mathbf{v}}(f) = \left[1 \; \boldsymbol{\alpha}^{T}\right], \tag{6.131}$$

from which we deduce the LCMV filter:

$$\widetilde{\mathbf{h}}_{\mathrm{LCMV}}(f) = \boldsymbol{\Phi}_{\mathbf{y}}^{-1}(f)\mathbf{C}_{\widetilde{X}\mathbf{v}}(f)\left[\mathbf{C}_{\widetilde{X}\mathbf{v}}^{H}(f)\boldsymbol{\Phi}_{\mathbf{y}}^{-1}(f)\mathbf{C}_{\widetilde{X}\mathbf{v}}(f)\right]^{-1}\begin{bmatrix} 1 \\ \boldsymbol{\alpha} \end{bmatrix}. \tag{6.132}$$

We see from (6.132) that we must have $Q + 1 \leq M$, otherwise the matrix $\mathbf{C}_{\widetilde{X}\mathbf{v}}^{H}(f)\boldsymbol{\Phi}_{\mathbf{y}}^{-1}(f)\mathbf{C}_{\widetilde{X}\mathbf{v}}(f)$ is not invertible. For $Q+1 > M$, the LCMV filter does not exist and for $Q + 1 = M$, the LCMV filter simplifies to

$$\widetilde{\mathbf{h}}_{\mathrm{LCMV}}(f) = \mathbf{C}_{\widetilde{X}\mathbf{v}}^{-H}(f)\begin{bmatrix} 1 \\ \boldsymbol{\alpha} \end{bmatrix}. \tag{6.133}$$

Finally, we see that the LCMV filtering matrix for the estimation of $\mathbf{x}(f)$ is

$$\begin{aligned}\mathbf{H}_{\mathrm{LCMV}}(f) = {} & \mathbf{q}_{\mathbf{x},1}(f)\left[1 \; \boldsymbol{\alpha}^{T}\right]\left[\mathbf{C}_{\widetilde{X}\mathbf{v}}^{H}(f)\boldsymbol{\Phi}_{\mathbf{y}}^{-1}(f)\mathbf{C}_{\widetilde{X}\mathbf{v}}(f)\right]^{-1} \\ & \times \mathbf{C}_{\widetilde{X}\mathbf{v}}^{H}(f)\boldsymbol{\Phi}_{\mathbf{y}}^{-1}(f). \end{aligned} \tag{6.134}$$

REFERENCES

[1] J. Benesty, J. Chen, Y. Huang, Microphone Array Signal Processing, Springer-Verlag, Berlin, Germany, 2008.

[2] J. Benesty, J. Chen, E. Habets, Speech Enhancement in the STFT Domain, Springer Briefs in Electrical and Computer Engineering, Springer-Verlag, 2011.

[3] J.P. Dmochowski, J. Benesty, Microphone arrays: fundamental concepts, in: I. Cohen, J. Benesty, S. Gannot (Eds.), Speech Processing in Modern Communication—Challenges and Perspectives, Springer-Verlag, Berlin, Germany, 2010, pp. 199–223 (Chapter 8).

[4] D.H. Johnson, D.E. Dudgeon, Array Signal Processing—Concepts and Techniques, Prentice-Hall, Englewood Cliffs, NJ, 1993.

[5] G.W. Elko, J. Meyer, Microphone arrays, in: J. Benesty, M.M. Sondhi, Y. Huang (Eds.), Springer Handbook of Speech Processing, Springer-Verlag, Berlin, Germany, 2008, pp. 1021–1041 (Chapter 48).

[6] J.N. Franklin, Matrix Theory, Prentice-Hall, Englewood Cliffs, NJ, 1968.

[7] J. Benesty, J. Chen, Y. Huang, I. Cohen, Noise Reduction in Speech Processing, Springer-Verlag, Berlin, Germany, 2009.

[8] Y. Ephraim, D. Mallah, Speech enhancement using a minimum mean-square error short-time spectral amplitude estimator, IEEE Trans. Acoust. Speech Signal Process. ASSP-32 (1984) 1109–1121.

[9] J. Capon, High resolution frequency-wavenumber spectrum analysis, Proc. IEEE 57 (1969) 1408–1418.

[10] R.T. Lacoss, Data adaptive spectral analysis methods, Geophysics 36 (1971) 661–675.

[11] M. Souden, J. Benesty, S. Affes, On the global output SNR of the parameterized frequency-domain multichannel noise reduction Wiener filter, IEEE Signal Process. Lett. 17 (2010) 425–428.

[12] O. Frost, An algorithm for linearly constrained adaptive array processing, Proc. IEEE 60 (1972) 926–935.

[13] M. Er, A. Cantoni, Derivative constraints for broad-band element space antenna array processors, IEEE Trans. Acoust. Speech Signal Process. 31 (1983) 1378–1393.

CHAPTER 7

A Bayesian Approach to the Speech Subspace Estimation

In all previous chapters, we showed the importance of the speech subspace; so it is extremely important in practice to find good ways to estimate it. One very promising possibility is the Bayesian approach based on the Stiefel manifold and the Bingham distribution. This chapter explores this avenue. We would like to point out that all the ideas presented in the following are borrowed from [1] but adapted to our specific problem of speech enhancement.

7.1 SIGNAL MODEL AND PROBLEM FORMULATION

In this chapter, we consider the signal model of Chapter 2, i.e.,

$$\mathbf{y} = \mathbf{T_x}\widetilde{\mathbf{x}} + \mathbf{v}. \tag{7.1}$$

To simplify the presentation, we assume that all signals are real; the generalization to the complex-valued case is straightforward if it is needed. We recall that $\mathbf{y}$ is the observation signal vector of length M, $\mathbf{T_x}$ is a semi-orthogonal matrix (i.e., $\mathbf{T_x}^T\mathbf{T_x} = \mathbf{I}_P$) of size $M \times P$, $\mathbf{T_x}\mathbf{T_x}^T$ is an orthogonal projection matrix, $\widetilde{\mathbf{x}}$ is the transformed desired signal vector, and $\mathbf{v}$ is the noise signal vector.

All optimal filtering matrices depend explicitly on $\mathbf{T_x}$. Therefore, $\mathbf{T_x}$ needs to be estimated. In speech enhancement applications, a rough estimate of this matrix, denoted by $\overline{\mathbf{T}}_\mathbf{x}$, can be obtained as follows. During silences, we estimate the correlation matrix of the noise, $\mathbf{\Phi_v}$, that we denote by $\overline{\mathbf{\Phi}}_\mathbf{v}$. It is easy to get an estimate of the correlation matrix of the observations, i.e., $\overline{\mathbf{\Phi}}_\mathbf{y}$. We then see that the estimate of the speech correlation matrix is $\overline{\mathbf{\Phi}}_\mathbf{x} = \overline{\mathbf{\Phi}}_\mathbf{y} - \overline{\mathbf{\Phi}}_\mathbf{v}$ and from its eigenvalue decomposition, we finally deduce $\overline{\mathbf{T}}_\mathbf{x}$. While this approach gives satisfactory results if the noise is stationary, it may not be the case for the nonstationary scenario. Therefore, it may be of interest to find better estimators for $\mathbf{T_x}$, which can be seen as the steering matrix corresponding to $\widetilde{\mathbf{x}}$. One very important way toward this direction is the Bayesian approach in which $\mathbf{T_x}$ is considered

Speech Enhancement. http://dx.doi.org/10.1016/B978-0-12-800139-4.00007-4

as random. This method is explored in the rest of this chapter. For that, we assume that $\mathbf{T_x}$ is assigned some prior distribution, $f(\mathbf{T_x})$, and our objective is to estimate $\mathbf{T_x}$ from the posterior distribution, $f(\mathbf{T_x} \mid \mathbf{y})$. We further assume that $\mathbf{T_x}$ and $\widetilde{\mathbf{x}}$ are independent.

7.2 ESTIMATION BASED ON THE MINIMUM MEAN-SQUARE DISTANCE

The Stiefel manifold [2] is the set of all semi-orthogonal matrices, $\mathbf{T_x}$. The classical MSE defined as

$$\begin{aligned} J_{\mathrm{MSE}}\left(\widehat{\mathbf{T}}_{\mathbf{x}}, \mathbf{T_x}\right) &= E\left(\left\|\widehat{\mathbf{T}}_{\mathbf{x}} - \mathbf{T_x}\right\|_{\mathrm{F}}^2\right) \\ &= E\left\{\mathrm{tr}\left[\left(\widehat{\mathbf{T}}_{\mathbf{x}} - \mathbf{T_x}\right)\left(\widehat{\mathbf{T}}_{\mathbf{x}} - \mathbf{T_x}\right)^T\right]\right\}, \end{aligned} \tag{7.2}$$

where $\widehat{\mathbf{T}}_{\mathbf{x}}$ is an estimate of $\mathbf{T_x}$, is not really appropriate on the Stiefel manifold while it is on the Euclidean space. The most natural metric on the Stiefel manifold is the mean-square distance (MSD) defined as [1–3]

$$\begin{aligned} J_{\mathrm{MSD}}\left(\widehat{\mathbf{T}}_{\mathbf{x}}, \mathbf{T_x}\right) &= E\left(\left\|\widehat{\mathbf{T}}_{\mathbf{x}}\widehat{\mathbf{T}}_{\mathbf{x}}^T - \mathbf{T_x}\mathbf{T_x}^T\right\|_{\mathrm{F}}^2\right) \\ &= E\left\{\mathrm{tr}\left[\left(\widehat{\mathbf{T}}_{\mathbf{x}}\widehat{\mathbf{T}}_{\mathbf{x}}^T - \mathbf{T_x}\mathbf{T_x}^T\right)\left(\widehat{\mathbf{T}}_{\mathbf{x}}\widehat{\mathbf{T}}_{\mathbf{x}}^T - \mathbf{T_x}\mathbf{T_x}^T\right)^T\right]\right\}. \end{aligned} \tag{7.3}$$

Indeed, the signal model in (7.1) can also be expressed as

$$\mathbf{y} = \mathbf{T_x}\mathbf{T_x}^T\mathbf{x} + \mathbf{v}. \tag{7.4}$$

We see clearly from the previous equation that we are more interested in the projection matrix $\mathbf{T_x}\mathbf{T_x}^T$ than in $\mathbf{T_x}$. As a consequence, it is more relevant to compare the two projection matrices $\mathbf{T_x}\mathbf{T_x}^T$ and $\widehat{\mathbf{T}}_{\mathbf{x}}\widehat{\mathbf{T}}_{\mathbf{x}}^T$ than the two semi-orthogonal matrices $\mathbf{T_x}$ and $\widehat{\mathbf{T}}_{\mathbf{x}}$.

By developing (7.3), we easily get

$$J_{\mathrm{MSD}}\left(\widehat{\mathbf{T}}_{\mathbf{x}}, \mathbf{T_x}\right) = 2P - 2E\left[\mathrm{tr}\left(\widehat{\mathbf{T}}_{\mathbf{x}}^T\mathbf{T_x}\mathbf{T_x}^T\widehat{\mathbf{T}}_{\mathbf{x}}\right)\right]. \tag{7.5}$$

As a result, the minimum MSD (MMSD) estimator of $\mathbf{T_x}$ is

$$\begin{aligned} \widehat{\mathbf{T}}_{\mathbf{x},\mathrm{MMSD}} &= \arg\min_{\widehat{\mathbf{T}}_{\mathbf{x}}} J_{\mathrm{MSD}}\left(\widehat{\mathbf{T}}_{\mathbf{x}}, \mathbf{T_x}\right) \\ &= \arg\max_{\widehat{\mathbf{T}}_{\mathbf{x}}} E\left[\mathrm{tr}\left(\widehat{\mathbf{T}}_{\mathbf{x}}^T\mathbf{T_x}\mathbf{T_x}^T\widehat{\mathbf{T}}_{\mathbf{x}}\right)\right]. \end{aligned} \tag{7.6}$$

Given the joint distribution $f\left(\mathbf{y}, \mathbf{T_x}\right)$, we have

$$E\left[\operatorname{tr}\left(\widehat{\mathbf{T}}_{\mathbf{x}}^{T}\mathbf{T_x}\mathbf{T}_{\mathbf{x}}^{T}\widehat{\mathbf{T}}_{\mathbf{x}}\right)\right] = \iint \operatorname{tr}\left(\widehat{\mathbf{T}}_{\mathbf{x}}^{T}\mathbf{T_x}\mathbf{T}_{\mathbf{x}}^{T}\widehat{\mathbf{T}}_{\mathbf{x}}\right) f\left(\mathbf{y}, \mathbf{T_x}\right) d\mathbf{y}\, d\mathbf{T_x}$$
$$= \int \left[\int \operatorname{tr}\left(\widehat{\mathbf{T}}_{\mathbf{x}}^{T}\mathbf{T_x}\mathbf{T}_{\mathbf{x}}^{T}\widehat{\mathbf{T}}_{\mathbf{x}}\right) f\left(\mathbf{T_x} \mid \mathbf{y}\right) d\mathbf{T_x}\right] f\left(\mathbf{y}\right) d\mathbf{y}. \tag{7.7}$$

It follows that

$$\widehat{\mathbf{T}}_{\mathbf{x},\mathrm{MMSD}} = \arg\max_{\widehat{\mathbf{T}}_{\mathbf{x}}} \int \operatorname{tr}\left(\widehat{\mathbf{T}}_{\mathbf{x}}^{T}\mathbf{T_x}\mathbf{T}_{\mathbf{x}}^{T}\widehat{\mathbf{T}}_{\mathbf{x}}\right) f\left(\mathbf{T_x} \mid \mathbf{y}\right) d\mathbf{T_x}$$
$$= \arg\max_{\widehat{\mathbf{T}}_{\mathbf{x}}} \operatorname{tr}\left\{\widehat{\mathbf{T}}_{\mathbf{x}}^{T}\left[\int \mathbf{T_x}\mathbf{T}_{\mathbf{x}}^{T} f\left(\mathbf{T_x} \mid \mathbf{y}\right) d\mathbf{T_x}\right]\widehat{\mathbf{T}}_{\mathbf{x}}\right\}$$
$$= \arg\max_{\widehat{\mathbf{T}}_{\mathbf{x}}} \operatorname{tr}\left[\widehat{\mathbf{T}}_{\mathbf{x}}^{T}\mathbf{M}\left(\mathbf{y}\right)\widehat{\mathbf{T}}_{\mathbf{x}}\right], \tag{7.8}$$

where

$$\mathbf{M}\left(\mathbf{y}\right) = \int \mathbf{T_x}\mathbf{T}_{\mathbf{x}}^{T} f\left(\mathbf{T_x} \mid \mathbf{y}\right) d\mathbf{T_x}. \tag{7.9}$$

Therefore, the MMSD estimate of $\mathbf{T_x}$ is given by the P largest eigenvectors of the matrix $\mathbf{M}\left(\mathbf{y}\right)$ [1], which we denote as

$$\widehat{\mathbf{T}}_{\mathbf{x},\mathrm{MMSD}} = \mathcal{P}_P\left[\mathbf{M}\left(\mathbf{y}\right)\right]. \tag{7.10}$$

The eigenvalue decomposition of $\mathbf{M}\left(\mathbf{y}\right)$ is

$$\mathbf{M}\left(\mathbf{y}\right) = \mathbf{Q}\left(\mathbf{y}\right)\boldsymbol{\Lambda}\left(\mathbf{y}\right)\mathbf{Q}^{T}\left(\mathbf{y}\right), \tag{7.11}$$

where

$$\boldsymbol{\Lambda}\left(\mathbf{y}\right) = \operatorname{diag}\left[\lambda_1\left(\mathbf{y}\right), \lambda_2\left(\mathbf{y}\right), \ldots, \lambda_M\left(\mathbf{y}\right)\right] \tag{7.12}$$

and $\lambda_1\left(\mathbf{y}\right) \geq \lambda_2\left(\mathbf{y}\right) \geq \cdots \geq \lambda_M\left(\mathbf{y}\right)$. We deduce that the average distance between $\widehat{\mathbf{T}}_{\mathbf{x},\mathrm{MMSD}}$ and $\mathbf{T_x}$ is given by [1,3]

$$J_{\mathrm{MSD}}\left(\widehat{\mathbf{T}}_{\mathbf{x},\mathrm{MMSD}}, \mathbf{T_x}\right)$$
$$= 2P - 2\int\left[\int \operatorname{tr}\left(\widehat{\mathbf{T}}_{\mathbf{x},\mathrm{MMSD}}^{T}\mathbf{T_x}\mathbf{T}_{\mathbf{x}}^{T}\widehat{\mathbf{T}}_{\mathbf{x},\mathrm{MMSD}}\right) f\left(\mathbf{T_x} \mid \mathbf{y}\right) d\mathbf{T_x}\right] f\left(\mathbf{y}\right) d\mathbf{y}$$
$$= 2P - 2\int \operatorname{tr}\left[\widehat{\mathbf{T}}_{\mathbf{x},\mathrm{MMSD}}^{T}\mathbf{M}\left(\mathbf{y}\right)\widehat{\mathbf{T}}_{\mathbf{x},\mathrm{MMSD}}^{T}\right] f\left(\mathbf{y}\right) d\mathbf{y}$$
$$= 2P - 2\sum_{p=1}^{P}\int \lambda_p\left(\mathbf{y}\right) f\left(\mathbf{y}\right) d\mathbf{y}. \tag{7.13}$$

7.3 A CLOSED-FORM SOLUTION BASED ON THE BINGHAM POSTERIOR

In the linear model shown in (7.1), we assume that the elements of $\mathbf{v}$ are independent and identically distributed zero-mean Gaussian noises with variance ϕ_v. Therefore, the probability density function (pdf) of $\mathbf{y}$, conditioned on $\mathbf{T_x}$ and $\widetilde{\mathbf{x}}$, is

$$f\left(\mathbf{y} \mid \mathbf{T_x}, \widetilde{\mathbf{x}}\right) \propto \operatorname{etr}\left[-\frac{1}{2\phi_v}\left(\mathbf{y} - \mathbf{T_x}\widetilde{\mathbf{x}}\right)^T\left(\mathbf{y} - \mathbf{T_x}\widetilde{\mathbf{x}}\right)\right], \tag{7.14}$$

where $\operatorname{etr}\left[\cdot\right]$ denotes the exponential of the trace of a matrix. To simplify the development, we assume that $\widetilde{\mathbf{x}}$ has a uniform prior distribution,[1] i.e., $f\left(\widetilde{\mathbf{x}}\right) = 1$. As a result,

$$\begin{aligned}
f\left(\mathbf{y} \mid \mathbf{T_x}\right) &= \int f\left(\mathbf{y} \mid \mathbf{T_x}, \widetilde{\mathbf{x}}\right) f\left(\widetilde{\mathbf{x}}\right) d\widetilde{\mathbf{x}} \\
&\propto \int \operatorname{etr}\left[-\frac{1}{2\phi_v}\left(\mathbf{y} - \mathbf{T_x}\widetilde{\mathbf{x}}\right)^T\left(\mathbf{y} - \mathbf{T_x}\widetilde{\mathbf{x}}\right)\right] d\widetilde{\mathbf{x}} \\
&\propto \operatorname{etr}\left(-\frac{1}{2\phi_v}\mathbf{y}^T\mathbf{y} + \frac{1}{2\phi_v}\mathbf{y}^T\mathbf{T_x}\mathbf{T_x}^T\mathbf{y}\right) \\
&\quad \times \int \operatorname{etr}\left[-\frac{1}{2\phi_v}\left(\widetilde{\mathbf{x}} - \mathbf{T_x}^T\mathbf{y}\right)^T\left(\widetilde{\mathbf{x}} - \mathbf{T_x}^T\mathbf{y}\right)\right] d\widetilde{\mathbf{x}} \\
&\propto \operatorname{etr}\left(-\frac{1}{2\phi_v}\mathbf{y}^T\mathbf{y} + \frac{1}{2\phi_v}\mathbf{y}^T\mathbf{T_x}\mathbf{T_x}^T\mathbf{y}\right).
\end{aligned} \tag{7.15}$$

It can be noticed how $f\left(\mathbf{y} \mid \mathbf{T_x}\right)$ depends on the projection matrix $\mathbf{T_x}\mathbf{T_x}^T$.

Since

$$\begin{aligned}
f\left(\mathbf{y}, \mathbf{T_x}\right) &= f\left(\mathbf{y} \mid \mathbf{T_x}\right) f\left(\mathbf{T_x}\right) \\
&= f\left(\mathbf{T_x} \mid \mathbf{y}\right) f\left(\mathbf{y}\right)
\end{aligned}$$

and

$$f\left(\mathbf{y}\right) = \int f\left(\mathbf{y} \mid \mathbf{T_x}\right) f\left(\mathbf{T_x}\right) d\mathbf{T_x}, \tag{7.16}$$

[1] Actually, a more appropriate prior for $\widetilde{\mathbf{x}}$ is the Laplace distribution but, with it, it will be extremely difficult to find a closed-form solution to our problem.

we obtain

$$f\left(\mathbf{T_x} \mid \mathbf{y}\right) \propto \frac{\operatorname{etr}\left(\frac{1}{2\phi_v}\mathbf{T}_\mathbf{x}^T\mathbf{y}\mathbf{y}^T\mathbf{T_x}\right) f\left(\mathbf{T_x}\right)}{\int \operatorname{etr}\left(\frac{1}{2\phi_v}\mathbf{T}_\mathbf{x}^T\mathbf{y}\mathbf{y}^T\mathbf{T_x}\right) f\left(\mathbf{T_x}\right) d\mathbf{T_x}} \propto \operatorname{etr}\left(\frac{1}{2\phi_v}\mathbf{T}_\mathbf{x}^T\mathbf{y}\mathbf{y}^T\mathbf{T_x}\right) f\left(\mathbf{T_x}\right). \tag{7.17}$$

Now, we need to find a relevant prior for $\mathbf{T_x}$.

One of the most widely accepted distributions on the Stiefel manifold is the Bingham distribution defined as

$$f_\mathrm{B}\left(\mathbf{T_x}\right) \propto \operatorname{etr}\left(\mathbf{T}_\mathbf{x}^T\mathbf{A}\mathbf{T_x}\right), \tag{7.18}$$

where $\mathbf{A}$ is an $M \times M$ symmetric matrix. We observe that the Bingham distribution depends also on the projection matrix $\mathbf{T_x}\mathbf{T}_\mathbf{x}^T$. It is important to introduce some knowledge about $\mathbf{T_x}$. For that, we can use the estimator $\overline{\mathbf{T}}_\mathbf{x}$ (see Section 7.1). In this case, (7.18) becomes

$$f_\mathrm{B}\left(\mathbf{T_x}\right) \propto \operatorname{etr}\left(\kappa\mathbf{T}_\mathbf{x}^T\overline{\mathbf{T}}_\mathbf{x}\overline{\mathbf{T}}_\mathbf{x}^T\mathbf{T_x}\right), \tag{7.19}$$

where κ is the concentration parameter. A large value of κ implies a high concentration of $\mathbf{T_x}$ around $\overline{\mathbf{T}}_\mathbf{x}$.

Substituting (7.19) into (7.17), we get

$$f\left(\mathbf{T_x} \mid \mathbf{y}\right) \propto \operatorname{etr}\left[\mathbf{T}_\mathbf{x}^T\left(\kappa\overline{\mathbf{T}}_\mathbf{x}\overline{\mathbf{T}}_\mathbf{x}^T + \frac{1}{2\phi_v}\mathbf{y}\mathbf{y}^T\right)\mathbf{T_x}\right], \tag{7.20}$$

which is recognized to be a Bingham distribution with parameter matrix:

$$\mathbf{A} = \kappa\overline{\mathbf{T}}_\mathbf{x}\overline{\mathbf{T}}_\mathbf{x}^T + \frac{1}{2\phi_v}\mathbf{y}\mathbf{y}^T. \tag{7.21}$$

It is shown in [1] that the eigenvectors of $\mathbf{M}\left(\mathbf{y}\right)$ [eq. (7.9)] coincide with those of $\mathbf{A}$ [eq. (7.21)]. Therefore,

$$\widehat{\mathbf{T}}_{\mathbf{x},\mathrm{MMSD}} = \mathcal{P}_P\left(\kappa\overline{\mathbf{T}}_\mathbf{x}\overline{\mathbf{T}}_\mathbf{x}^T + \frac{1}{2\phi_v}\mathbf{y}\mathbf{y}^T\right). \tag{7.22}$$

This result is extremely simple and makes sense intuitively since $\widehat{\mathbf{T}}_{\mathbf{x},\mathrm{MMSD}}$ depends on the available estimator and a correction term based on the observations. Also, it is easy to verify that the MMSD and maximum a posteriori (MAP) estimators are the same.

REFERENCES

[1] O. Besson, N. Dobigeon, J.-Y. Tourneret, Mnimum mean square distance estimation of a subspace, IEEE Trans. Signal Process. 59 (2011) 5709–5720.

[2] A. Edelman, T. Arias, S. Smith, The geometry of algorithms with orthogonality constraints, SIAM J. Matrix Anal. Appl. 20 (2) (1998) 303–353.

[3] A. Srivastava, A Bayesian approach to geometric subspace estimation, IEEE Trans. Signal Process. 48 (2000) 1390–1400.

CHAPTER 8

Evaluation of the Time-Domain Speech Enhancement Filters

So far, all the chapters have been dealing with the derivation of optimal filters for speech enhancement from a subspace perspective. In this chapter, we present evaluations of the time-domain versions of these filters for both single-channel as well as multichannel scenarios. First, we evaluate the single-channel scenario, where we consider both the case where the rank of the correlation matrix of the desired signal is less than the filter length and the case where it is full rank. Then, we proceed to evaluate the time-domain multichannel filters.

8.1 EVALUATION OF SINGLE-CHANNEL FILTERS

In this section, we evaluate the single-channel filters presented in Chapter 4 in terms of output SNR, signal reduction, and, when relevant, output signal-to-interference ratio (SIR).

8.1.1 Rank-Deficient Speech Correlation Matrix

First, we evaluate the filters presented in Section 4.4, which are derived under the assumption that the rank of the desired signal correlation matrix, $\mathbf{R}_\mathbf{x}$, is less than the filter length, L. For these evaluations, we assumed the desired signal to be periodic, since this enables us to model the statistics appearing in the filter design as well as in the performance measure expressions. In this way, we evaluate the filters without having to estimate any statistics, which insures a clear picture of the theoretical noise reduction and signal distortion properties of the filters. Another motivation for choosing periodic signals for this evaluation is the fact that most parts of speech signals can be considered as quasi-periodic.

More specifically, we used a periodic signal constituted by $C = 8$ harmonics (meaning that $P = 2C = 16$) with unit amplitudes (i.e., $|\alpha_c| = 1$ for $c = 1, \ldots, C$) and random phases as the desired signal in this experiment. The correlation matrix for the desired signal can then be

Speech Enhancement. http://dx.doi.org/10.1016/B978-0-12-800139-4.00008-6

modeled by [1]

$$\mathbf{R_x} = \bar{\mathbf{Z}}\bar{\mathbf{P}}\bar{\mathbf{Z}}^H, \tag{8.1}$$

where

$$\bar{\mathbf{Z}} = \begin{bmatrix} \mathbf{Z} \ \mathbf{Z}^* \end{bmatrix}, \tag{8.2}$$

$$\bar{\mathbf{P}} = \begin{bmatrix} \mathbf{P} & \mathbf{0}_{C\times C} \\ \mathbf{0}_{C\times C} & \mathbf{P} \end{bmatrix}, \tag{8.3}$$

with

$$\mathbf{Z} = \begin{bmatrix} \mathbf{z}(\omega_0) & \mathbf{z}(2\omega_0) & \dots & \mathbf{z}(C\omega_0) \end{bmatrix}, \tag{8.4}$$

$$\mathbf{z}(c\omega_0) = \begin{bmatrix} 1 & e^{-jc\omega_0} & \dots & e^{-jc\omega_0(L-1)} \end{bmatrix}^T, \tag{8.5}$$

$$\mathbf{P} = \text{diag}\left\{ \begin{bmatrix} |\alpha_1|^2 & \dots & |\alpha_C| \end{bmatrix}^T \right\}, \tag{8.6}$$

$(\cdot)^*$ denotes the elementwise complex conjugate of a matrix, ω_0 is the fundamental frequency of the periodic signal, C is the number of harmonics constituting the periodic signal, diag$\{\cdot\}$ denotes the transformation of a vector into a diagonal matrix, where the vector appears on the diagonal, and α_c denotes the complex amplitude of the cth harmonic.

We then assumed that the desired signal was corrupted by white Gaussian noise for our first evaluations such that the correlation matrix of the noise was given by $\mathbf{R_v} = \sigma_v^2 \mathbf{I}_L$, where σ_v^2 is the noise variance. The maximum SNR ($\mathbf{H}_{\text{max}}$), Wiener ($\mathbf{H}_{\text{W}}$), MVDR ($\mathbf{H}_{\text{M}}$), and tradeoff ($\mathbf{H}_{\text{T}}$) filters presented in Section 4.4 were then evaluated for this signal and noise scenario for different filter setups. Unless otherwise stated, the input SNR was 10 dB, the filter length was $L = 60$, and the tradeoff filter tuning parameter was $\mu = 10$. For each filter setup, 100 Monte-Carlo simulations were run, where the fundamental frequency was randomized by sampling from the uniform distribution $\mathcal{U}(0.05, 0.2)$, and the depicted performance measures were obtained as the average of the performance measures over the different Monte-Carlo trials.

First, we evaluated the output SNRs and the signal reduction factors of the filters versus different filter lengths. The results from this experiment are depicted in Fig. 8.1. As expected, this figure shows that the maximum

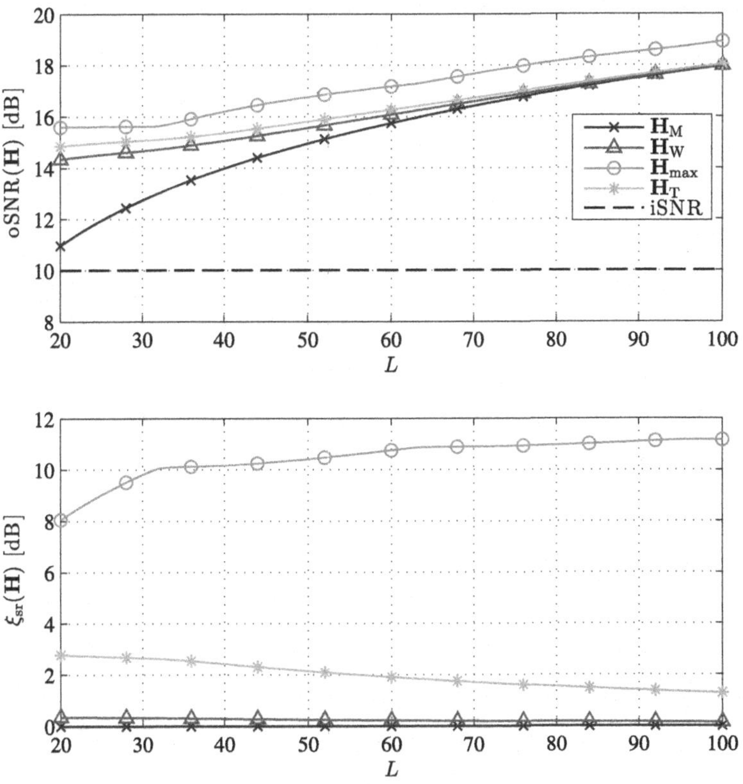

Fig. 8.1 Performance of the single-channel filters presented in Section 4.4 versus the filter length with white Gaussian noise.

SNR filter achieves the highest output SNR followed by the tradeoff filter, the Wiener filter, and, finally, the MVDR filter. The output SNR of the tradeoff filter is, however, dependent on the tuning parameter, and can be lowered and raised by decreasing or increasing the value of this parameter as shown in a later experiment. For increasing filter lengths, all of the filters achieve an increase in output SNR compared to the input SNR. Regarding distortion, which is measured through the signal reduction factor in this case, the MVDR filter is undistorted as expected, since its signal reduction factor equals one in all cases. The maximum SNR filter has the highest signal reduction factor, and it is even increasing for an increasing filter length, whereas the signal reduction factors of the tradeoff and Wiener filters decrease toward one for increasing filter lengths. In the next experiment, we evaluated the filters' performance

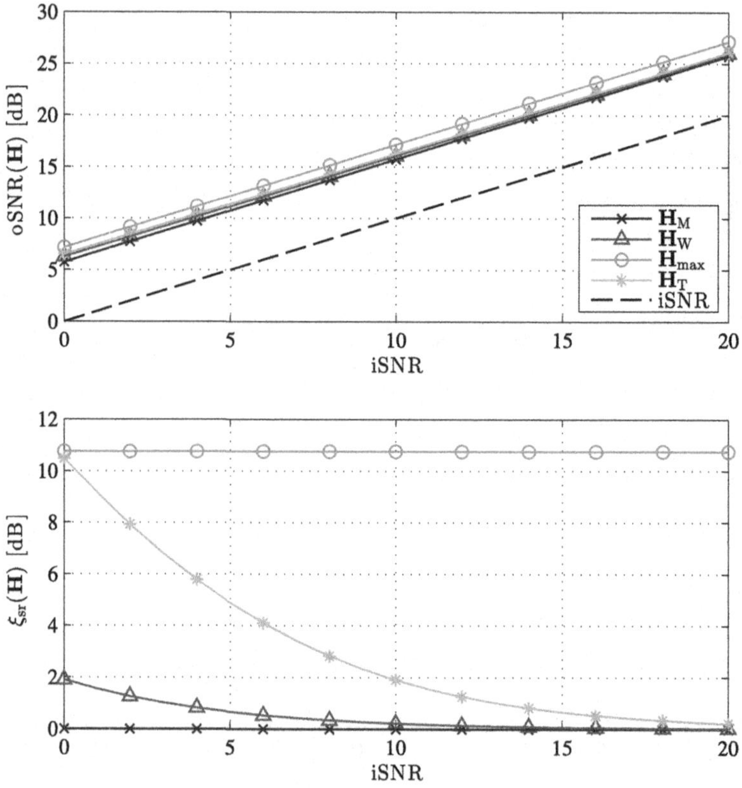

Fig. 8.2 Performance of the single-channel filters presented in Section 4.4 versus the input SNR with white Gaussian noise.

versus the input SNR, and the results are depicted in Fig. 8.2. The figure shows the relative performance between the filters in terms of output SNR as in the previous experiment, and the gain in SNR is nearly constant for different input SNRs. The signal reduction factor of the maximum SNR filter is also constant for different input SNRs, while the signal reduction factors of the Wiener and tradeoff filters decrease toward one for increasing input SNR. Again, we observe that the MVDR filter has no distortion. At the final experiment for this signal and noise scenario, we evaluated the filters for different μ's. The outcomes of this experiment can be seen in Fig. 8.3. The interesting part of these plots are the performances of the tradeoff filter, which is dependent on μ. In general, the output SNR of the tradeoff filter increases when μ is increased and so is the signal reduction factor. In other words, the gain in noise reduction comes at the

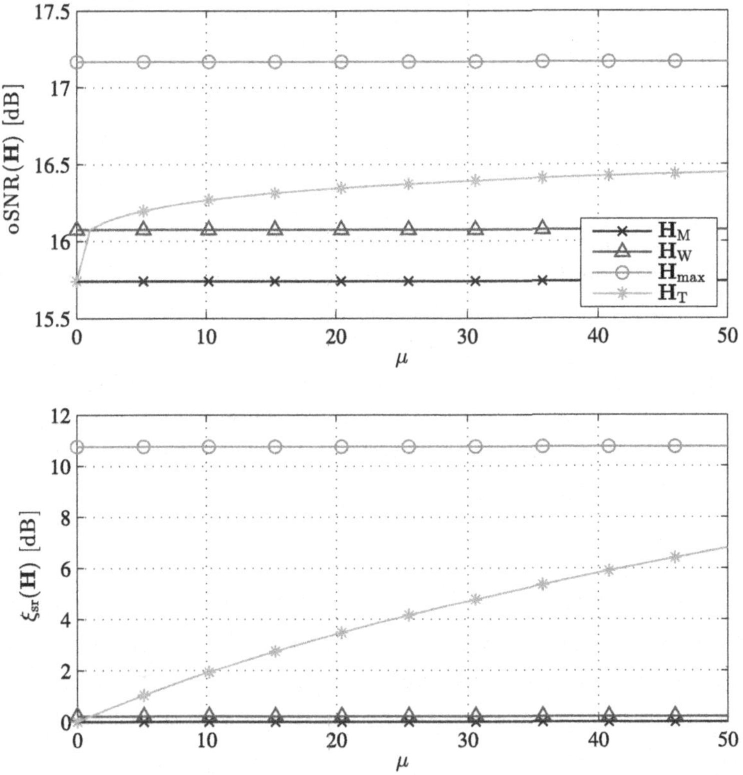

Fig. 8.3 Performance of the single-channel filters presented in Section 4.4 versus the tradeoff parameter with white Gaussian noise.

cost of an increased signal distortion. We note that for $\mu = 0$, the tradeoff filter resembles the MVDR filter, while it resembles the Wiener filter for $\mu = 1$. Moreover, we note that the output saturates for increasing μ, and that we do not attain the maximum output SNR for any μ.

Then, we considered a different signal and noise scenario, where the noise was constituted by the sum of an interfering noise source and white Gaussian noise. In this scenario, unless otherwise stated, the desired signal was again a periodic signal, but here it was constituted by $C = 6$ unit amplitude harmonics, and the fundamental frequency of the signal was randomized in series of Monte-Carlo simulations as in the previous experiments. The interferer was a single, real sinusoid with a frequency of approximately 1.2 times the fundamental frequency of the desired signal. The input SNR, which in this case is defined as the ratio between the powers of the desired signal and the white noise, was 10 dB, while the input SIR was 0 dB. In this noise scenario, we can model the noise correlation matrix as

$$\mathbf{R}_{\mathrm{v}} = \left[\mathbf{z}(1.2\omega_0)\ \mathbf{z}^*(1.2\omega_0)\right]\begin{bmatrix} P_{\mathrm{i}} & 0 \\ 0 & P_{\mathrm{i}} \end{bmatrix}\begin{bmatrix} \mathbf{z}^H(1.2\omega_0) \\ \mathbf{z}^T(1.2\omega_0) \end{bmatrix} + \sigma_v^2\mathbf{I}_L, \tag{8.7}$$

where P_{i} is the power of complex sinusoids constituting the real interfering, periodic signal. Again, the filter order was $L = 60$ except where indicated differently. With this setup, we evaluated the maximum SNR, Wiener, MVDR, tradeoff, and LCMV ($\mathbf{H}_{\mathrm{L}}$) filters in terms of output SNRs, signal reduction factors, and output SIRs. First, these measures were evaluated versus the filter length as depicted in Fig. 8.4. We see that the relations between the output SNRs and signal reduction factors of the maximum SNR, Wiener, MVDR, and tradeoff filters are similar to the relations in the corresponding experiment with white noise only (Fig. 8.1). The output SNR of the LCMV filter, on the other hand, is lower than those of the other filters for all filter lengths, while the difference in output SNRs become smaller as we increase the filter length. This is expected, as the LCMV filter has less degrees of freedom for broadband noise reduction. Due to its constraints, the LCMV filter is distortionless as the MVDR filter, which is also confirmed from these results. When it comes to output SIR, the LCMV filter outperforms the other filters as expected due to its null constraints. The maximum SNR filter also achieves a high output SIR compared to the remaining filters in this experiment. We also evaluated the filters as a function of the input SNR as showed in Fig. 8.5. Again, the relation between the filters, except for the LCMV filter, is similar to what was observed in the previous experiments (Fig. 8.2). For an increasing input SNR, we observe that the performances of the Wiener, MVDR, LCMV filters become close. Moreover, we observe that the LCMV filter outperforms the other filters in terms of output SIR, and that its output SIR is independent on the input SNR.

8.1.2 Full-Rank Speech Correlation Matrix

We then proceed to evaluate the filters presented in Section 4.5, which are applicable even when the desired signal correlation matrix has full rank. This is typically the case in practice for unvoiced speech, which has a broadband, noise-like spectrum. For these experiments, we therefore used a signal generated by an autoregressive (AR) process as the desired

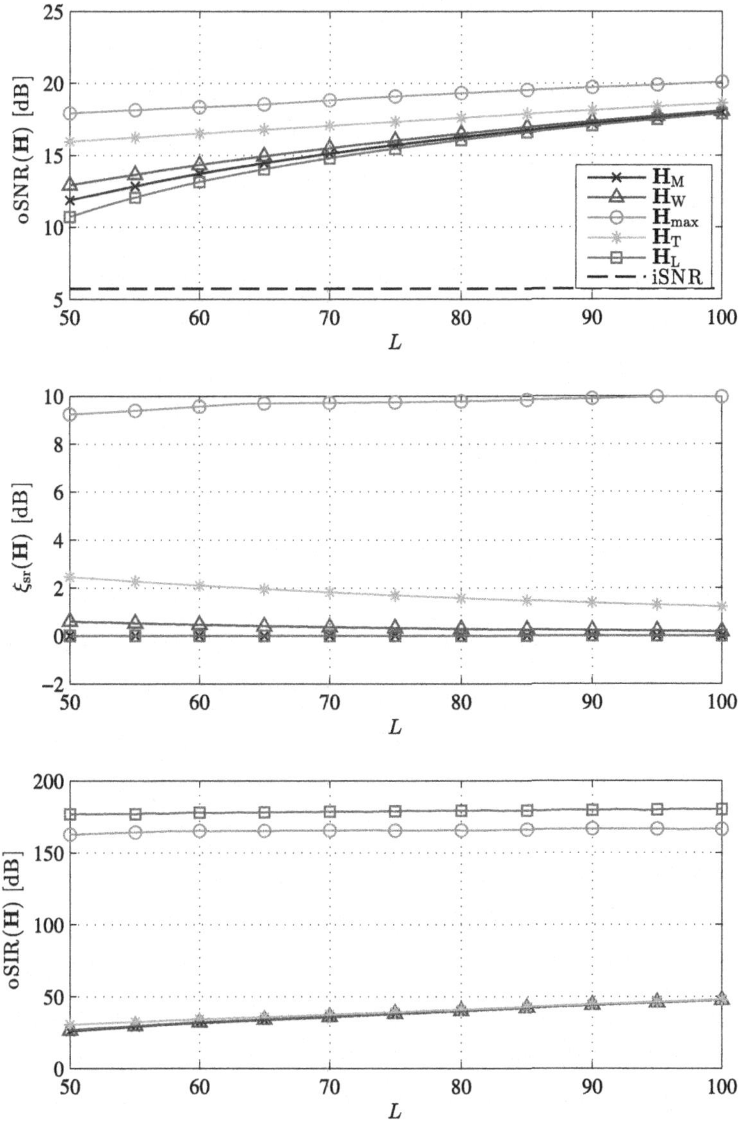

Fig. 8.4 Performance of the single-channel filters presented in Section 4.4 versus the filter length with white Gaussian noise and an interfering, periodic source.

signal, which is widely used to model unvoiced speech. More specifically, we used an AR(2) process, for which we have [2]

$$r_x(0) = \frac{1}{1 - \phi_1 \rho_x(1) - \phi_2 \rho_x(2)}, \tag{8.8}$$

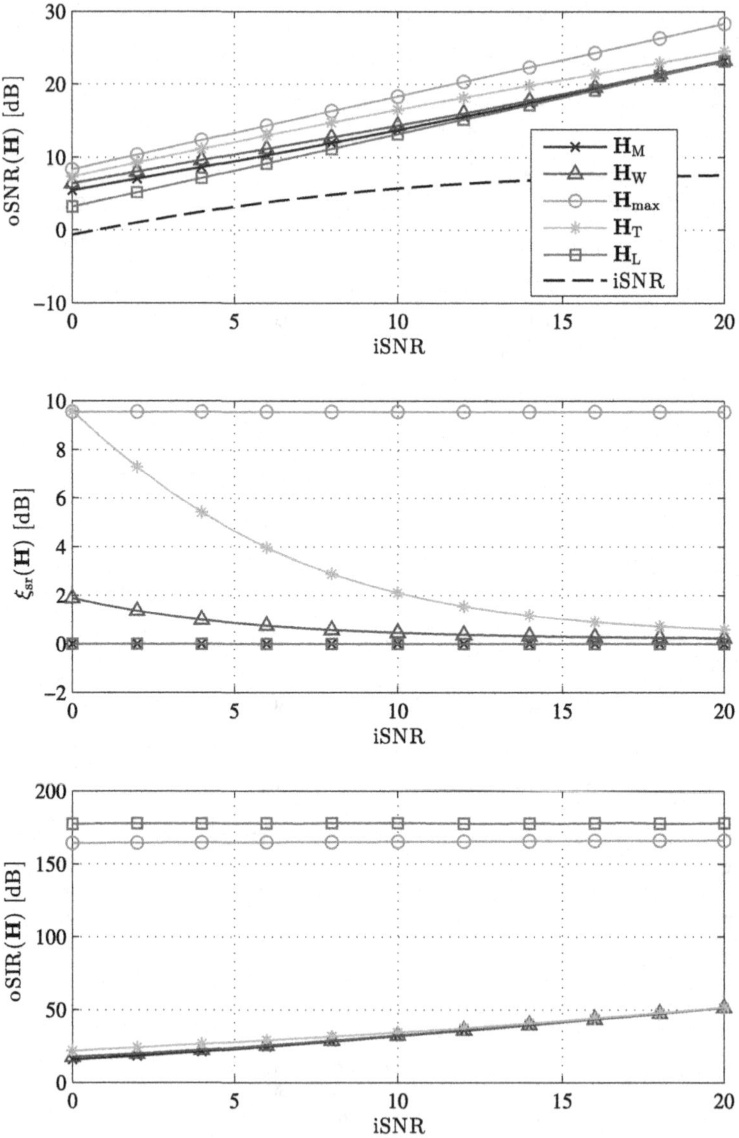

Fig. 8.5 Performance of the single-channel filters presented in Section 4.4 versus the input SNR with white Gaussian noise and an interfering, periodic source.

$$r_x(1) = \frac{\phi_1}{1 - \phi_2} r_x(0), \tag{8.9}$$

$$r_x(\kappa) = \phi_1 r_x(\kappa - 1) + \phi_2 r_x(\kappa - 2), \quad \text{for } \kappa \geq 2, \tag{8.10}$$

where $r_x(\kappa) = E[x(k)x(k+\kappa)]$ is the autocorrelation function for $x(k)$, $\rho_x(\kappa) = r_x(\kappa)/r_x(0)$ is the normalized autocorrelation function for $x(k)$, and ϕ_i denotes the ith AR parameter. Furthermore, it can be shown that

$$\rho_x(1) = \frac{\phi_1}{1-\phi_2}, \quad (8.11)$$

$$\rho_x(2) = \frac{\phi_1^2 + \phi_2(1-\phi_2)}{1-\phi_2}. \quad (8.12)$$

Utilizing the expressions (8.8)–(8.12), we have a closed-form expression for the correlation matrix of the desired signal, which is used to design the different filters in the evaluations of this subsection.

As in Section 8.1.1, we conducted 100 Monte-Carlo trials for different parameter settings in these experiments. The AR parameters were randomized in each of these trials, and they were found by fitting an AR(2) process to 500 samples of a periodic signal with $C=8$ harmonics with random phases, the real amplitudes $|\boldsymbol{\alpha}| = [0.8, 1, 0.8, 0.6, 0.6, 0.4, 0.1, 0.05]$, and a pitch sampled from $\mathcal{U}(0.05, 0.2)$. The noise was again assumed to be white Gaussian with correlation matrix $\mathbf{R}_\mathbf{v} = \sigma_v^2\mathbf{I}_L$. The maximum SNR ($\mathbf{H}_{\text{max}}$), Wiener ($\mathbf{H}_\text{W}$), MVDR ($\mathbf{H}_\text{M}$), and tradeoff ($\mathbf{H}_\text{T}$) filters presented in Section 4.5 were then evaluated for this signal and noise scenario for different filter setups. Unless stated otherwise, the ratio between the powers of the desired signal and the white noise was 10 dB, the filter length was $L = 60$, the tradeoff filter tuning parameter was $\mu = 10$, and the estimation of the $P = 10$ first samples of the desired signal vector was considered.

With this setup, we first evaluated the filters' performance for different filter lengths, L. The resulting plots are found in Fig. 8.6. The first thing we observe from the figure is that the input SNR decreases for increasing filter lengths, which at first glance might be counterintuitive since the ratio between the powers of the desired signal and the white noise is constant in all simulations, but it is a consequence of the input SNR definition in (4.85) for the filters in Section 4.5. For these filters, the noise terms are defined as the sum of the real noise and a so-called interference term. The power of the interference term increases for increasing sample lengths, hence the input SNR will be decreasing for increasing filter lengths. Next, we observe that the maximum SNR filter achieves

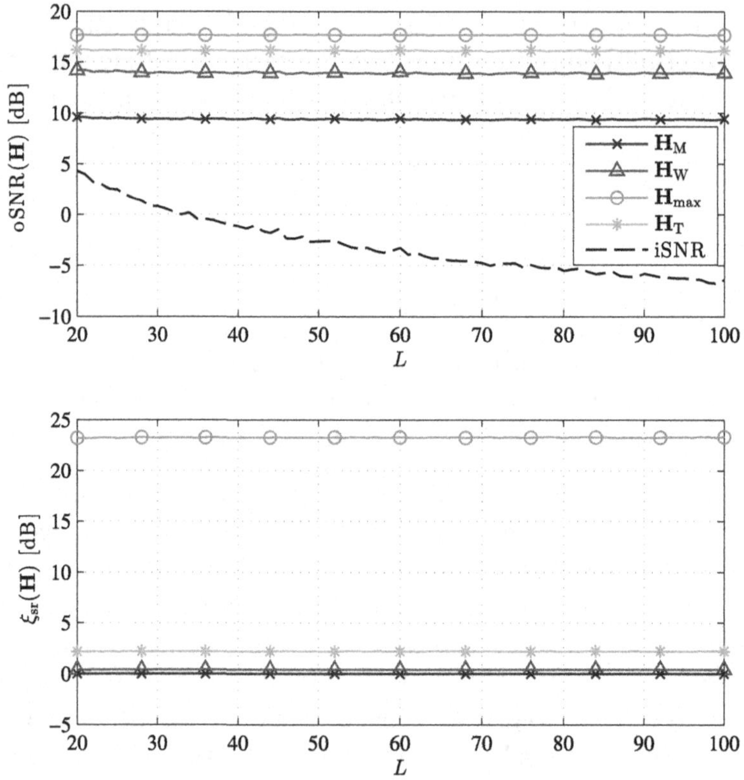

Fig. 8.6 Performance of the single-channel filters presented in Section 4.5 versus the filter length with white Gaussian noise.

the highest output SNR, as expected, followed by the tradeoff ($\mu = 10$), Wiener, and MVDR filters. In general, the output SNRs of the filters are constant for increasing filter lengths, whereas their SNR gain in relation to the input is increasing due to the decreasing input SNR. From the measured signal reduction factors, we see that the highest and lowest signal distortion is achieved with the maximum SNR and MVDR filters, respectively. We also observe that the higher output SNR of the tradeoff filter in comparison with the Wiener and MVDR filters comes at the cost of a higher signal distortion.

We also evaluated the performance versus the number of signal samples to be estimated, P, and the results are provided in Fig. 8.7. From this figure, we make the interesting observation, that the SNR gain is largest for a relatively small P. In general, the SNR gain is highest for all filters

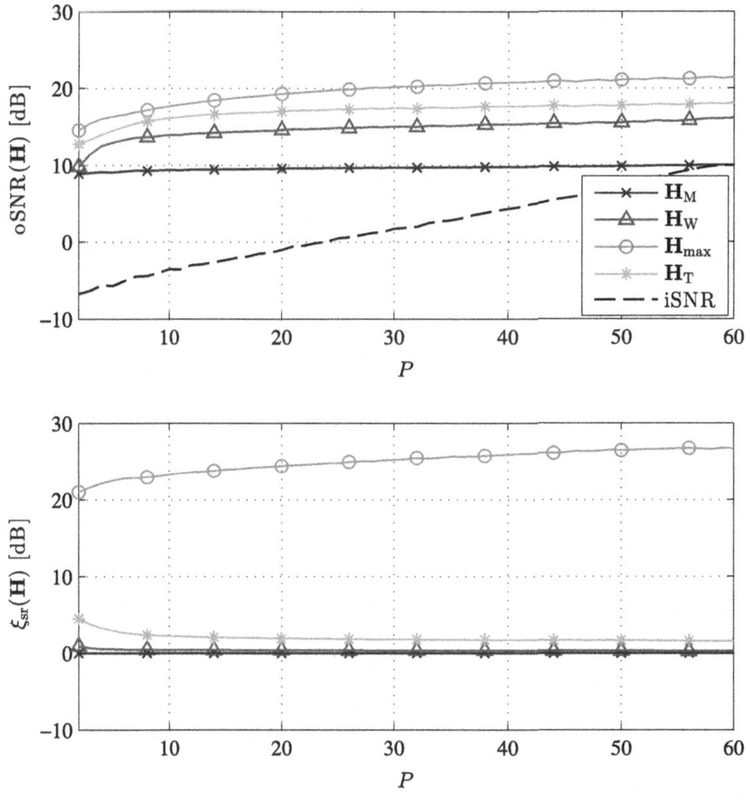

Fig. 8.7 Performance of the single-channel filters presented in Section 4.5 versus different number of samples to be estimated with white Gaussian noise.

in the evaluation for $1 \leq P \leq 10$, whereas the SNR gain is decreasing for higher P's. The relationship between the output SNRs and signal reduction factors of the filters is the same as in the evaluation versus the filter length. However, it is interesting to see that the signal reduction factor, i.e., the signal distortion, of the tradeoff filter can be decreased by estimating more samples simultaneously. Third, the performance investigation was conducted versus different input SNRs, as reported in Fig. 8.8. It can be seen that the output SNR of the maximum SNR filter is linearly increasing and that it has a constant SNR gain. The Wiener, MVDR, and tradeoff filters also have increasing output SNRs, but their SNR gains are decreasing for increasing input SNR. On the other hand, the signal reduction factors of the Wiener and tradeoff filters are decreasing for increasing input SNR, as opposed to the signal reduction factor for

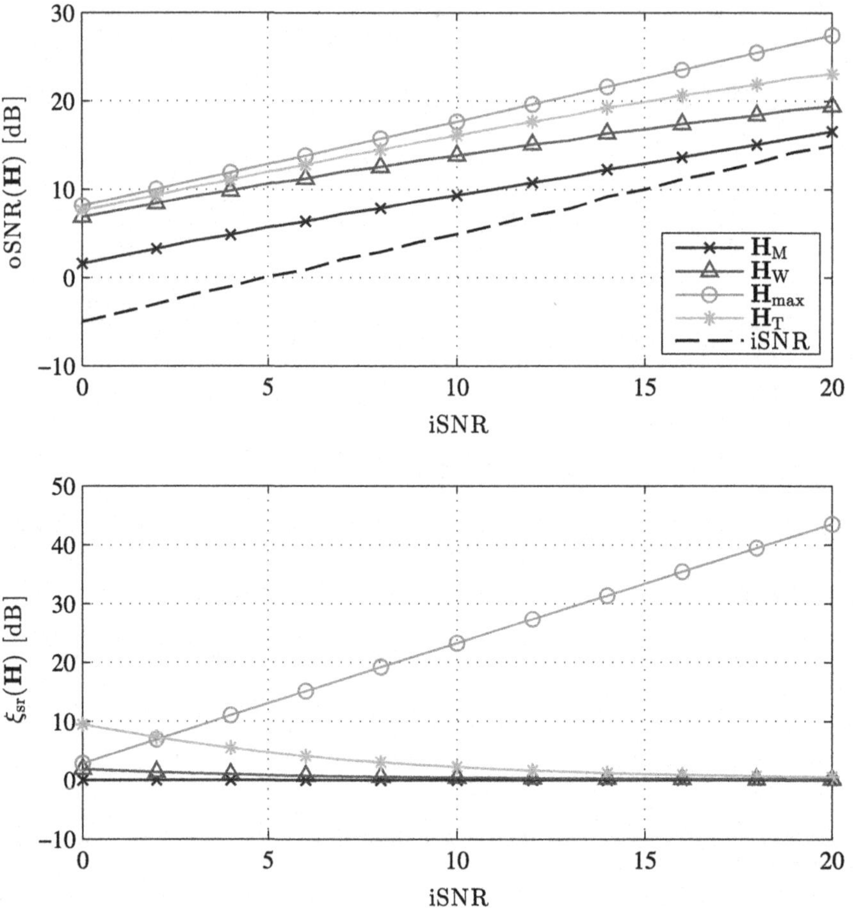

Fig. 8.8 Performance of the single-channel filters presented in Section 4.5 versus the input SNR with white Gaussian noise.

the maximum SNR filter which is increasing. The signal reduction factor of the MVDR filter is of course one as expected. Finally, the results in Fig. 8.9 were obtained in evaluations where the tradeoff parameter, μ, was altered. The performance measures of the maximum SNR, Wiener, and MVDR filters are of course unaffected for different μ's, while the output SNR and signal reduction factor of the tradeoff filter is increasing and decreasing, respectively, for increasing μ's. It is confirmed that for $\mu = 0$, the tradeoff filter resembles the MVDR filter, and it resembles the Wiener filter for $\mu = 1$.

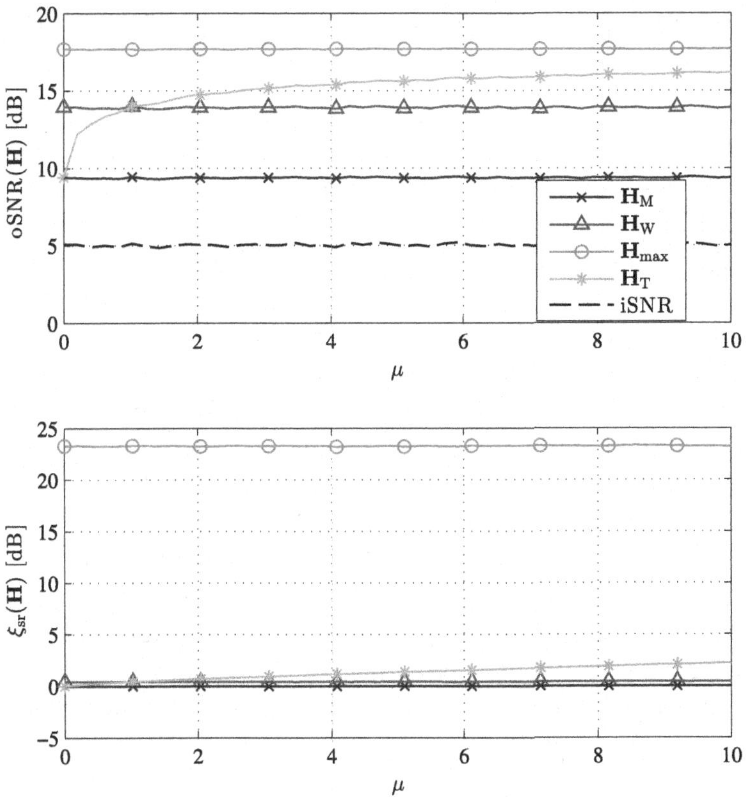

Fig. 8.9 Performance of the single-channel filters presented in Section 4.5 versus the tradeoff parameter with white Gaussian noise.

We then proceed to consider another signal and noise scenario, where an interfering, periodic noise source was added in addition to the white Gaussian noise. The interfering, periodic source was constituted by $C_\mathrm{i} = 1$ interfering sinusoid with a frequency of $\omega_\mathrm{i} = \omega_0 + 0.2$, with ω_0 being the fundamental frequency of the periodic signal used for generating the AR parameters. The power of the periodic interferer was scaled such that the ratio between the powers of the desired signal and the periodic interferer was 0 dB. As recommended by the results in Fig. 8.7, a relatively small number of samples to be estimated was chosen, i.e., $P = 10$. Unless stated otherwise, the filter length was set to $L = 60$, and the ratio between the powers of the desired signal and the white noise was 10 dB. With this setup, we then evaluated the maximum SNR, Wiener, MVDR, tradeoff, and LCMV ($\mathbf{H}_L$) filters in terms of output SNRs, signal reduction factors, and output SIRs.

First, this evaluation was conducted for different filter lengths, with the results being presented in Fig. 8.10. All the filters have slightly increasing output SNRs for increasing filter lengths, and the maximum SNR filter expectedly has the highest output SNR followed by the tradeoff, Wiener, MVDR, and LCMV filters. While the MVDR and LCMV filters have the lowest output SNRs, they are distortionless as opposed to the other filters. The increased output SNR of the tradeoff filter comes at the cost of an increased signal reduction factor. Finally, we observe that the LCMV filter might still be preferred despite its low output SNR in some scenarios, since it clearly outperforms the other filters in terms of output SIR, with an output SIR approaching infinity. In the following evaluation, the filters were then evaluated for different input SNRs as depicted in Fig. 8.11. Here, we see that the output SNR of the maximum SNR filter increases at a higher rate than for the other filters as a function of the input SNR. However, the maximum SNR filter also has an increasing signal reduction factor for increasing input SNR, whereas the signal reduction factor is decreasing for the other filters. The highest output SIR is achieved by the LCMV filter as expected followed by the maximum SNR filter.

8.2 EVALUATION OF MULTICHANNEL FILTERS

We also evaluated the multichannel filters presented in Chapter 5, and the results are presented in this section. In the multichannel scenario with reverberation, closed-form expressions of the correlation matrices of the desired signal and interfering sources for the filters' design are not directly available and difficult to obtain, so for these experiments, we generated synthetic, speech-like multichannel signals and estimated their statistics. The filters were then designed and evaluated using these statistics estimates. The desired signal was assumed to be generated from an AR(20) process. First, we generated the AR parameters by, again, fitting an AR(20) process to 500 samples of a periodic signal with $C = 8$ harmonics with amplitudes $|\boldsymbol{\alpha}| = [0.8, 1, 0.8, 0.6, 0.6, 0.4, 0.1, 0.05]$ and a fundamental frequency sampled from $\mathcal{U}(0.05, 0.2)$. Then, a white Gaussian noise signal of length $3LM$ was filtered through an IIR filter with

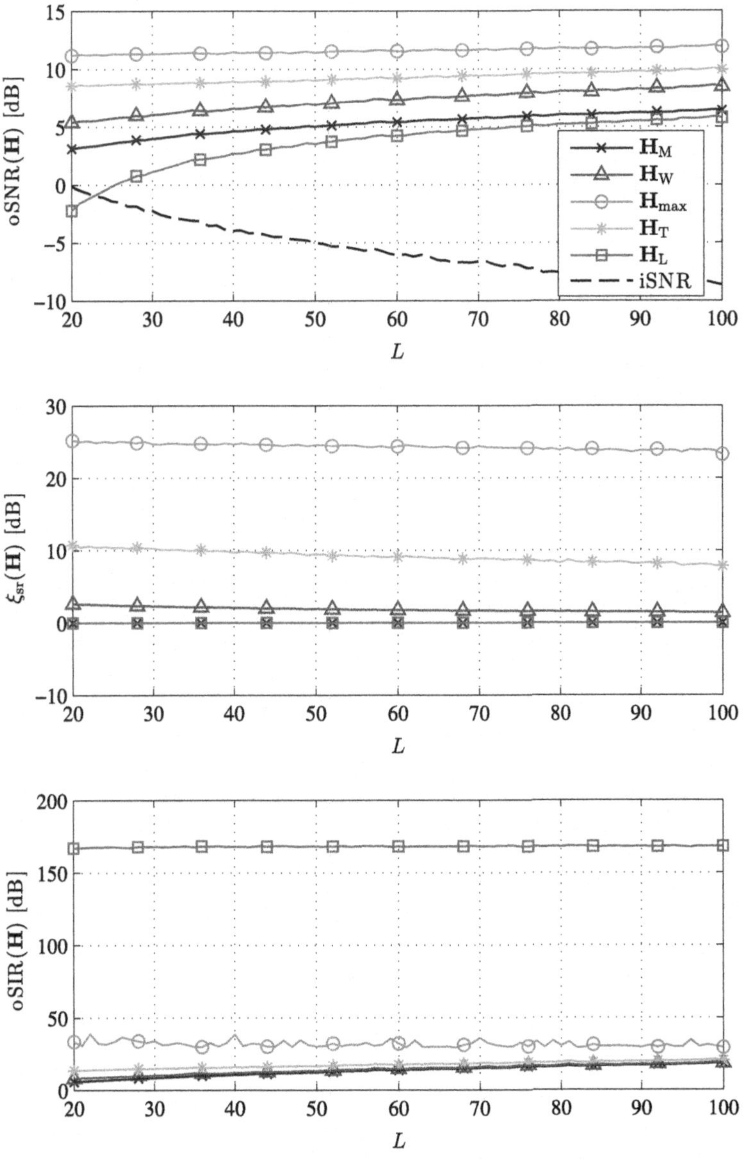

Fig. 8.10 Performance of the single-channel filters presented in Section 4.5 versus the filter length with white Gaussian noise and an interfering, periodic source.

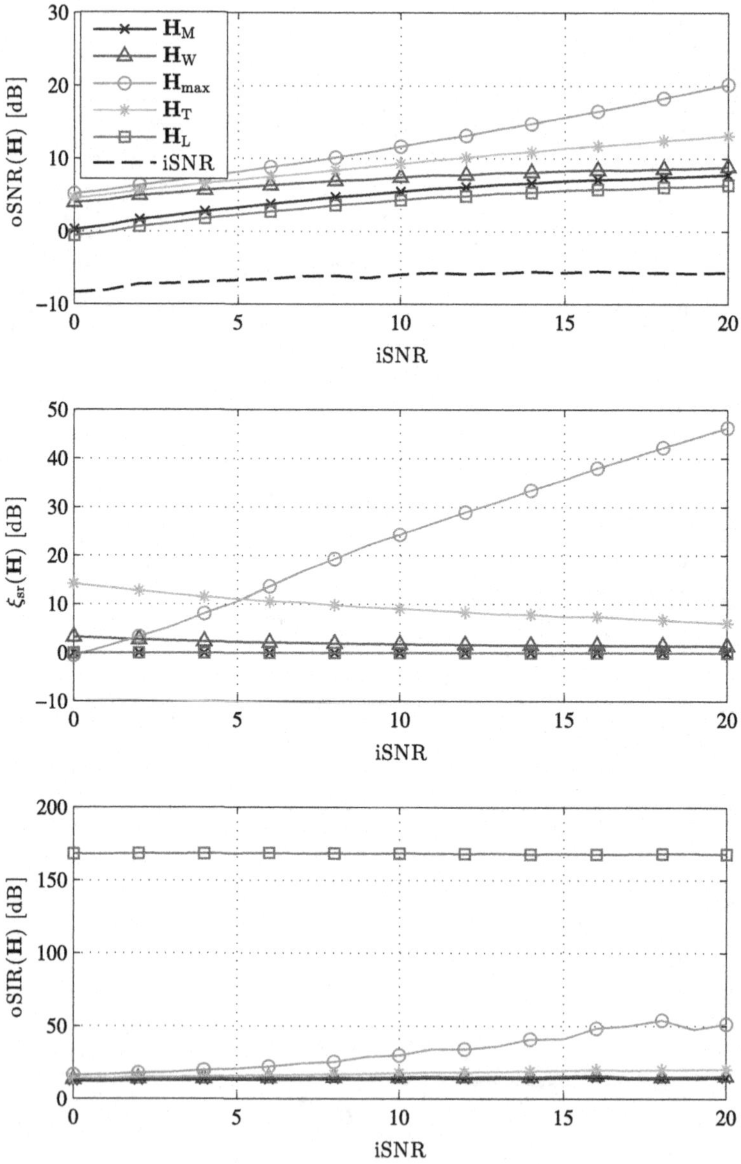

Fig. 8.11 Performance of the single-channel filters presented in Section 4.5 versus the input SNR with white Gaussian noise and an interfering, periodic source.

coefficients corresponding to the AR(20) parameters, to generate a single-channel signal.

To obtain the desired multichannel signal, we used an online available room impulse response (RIR) generator [3]. The RIRs were generated with the following setup: the sound velocity was 340 m/s, the sampling frequency was 8 kHz, the room was $5\times4\times3$ m and had a reverberation time of 0.4 s, the generated RIRs were 2048 samples long, the microphones were omnidirectional, the maximum number of reflections was used, and highpass filtering was enabled. Moreover, RIRs were generated for 10 microphones organized in a uniform linear array (ULA) structure with an intermicrophone spacing of 2 cm. The ULA was oriented perpendicular to the desired source. To evaluate more than one room setup, we considered different placements of the center of the ULA on a circle around the source position. More specifically, sets of RIRs were generated for 25 uniformly sampled points on this circle with a radius of 1 m. Then, the desired, synthetic multichannel signal was generated by filtering the single-channel AR(20) signal with the generated RIRs.

To generate the observed signal, we assumed the noise to be additive white Gaussian noise on each channel and uncorrelated across channels. The noise correlation matrix was therefore given by $\mathbf{R}_{\mathbf{v}} = \sigma_v^2 \mathbf{I}_{ML}$. For this signal and noise scenario, we then evaluated the maximum SNR, Wiener, MVDR, and tradeoff, multichannel filters presented in Section 5.4. Unless otherwise stated, the input SNR at the first microphone was 10 dB, the temporal filter length was $L = 30$, the number of microphones[1] was $M = 10$, and the tradeoff parameter was $\mu = 10$. For each simulation setup, we then ran 100 Monte-Carlo trials, and, in each trial, the ULA center position and the additive noise were randomized.

In the first evaluation, we investigated the filters' performance versus the temporal filter length, L, and the results are found in Fig. 8.12. As for the single-channel filters, the multichannel filters are ranked as: maximum SNR, tradeoff, Wiener, and MVDR in terms of highest output SNR. All filters have slightly increasing output SNRs for increasing filter lengths. While the MVDR has the lowest output SNR, it outperforms the other filters (although only by a small margin for the Wiener filter) in terms of

[1]Note that RIRs were generated for 10 microphones, so, in the simulations with less microphones, the M central RIRs of the 10-element ULA were used.

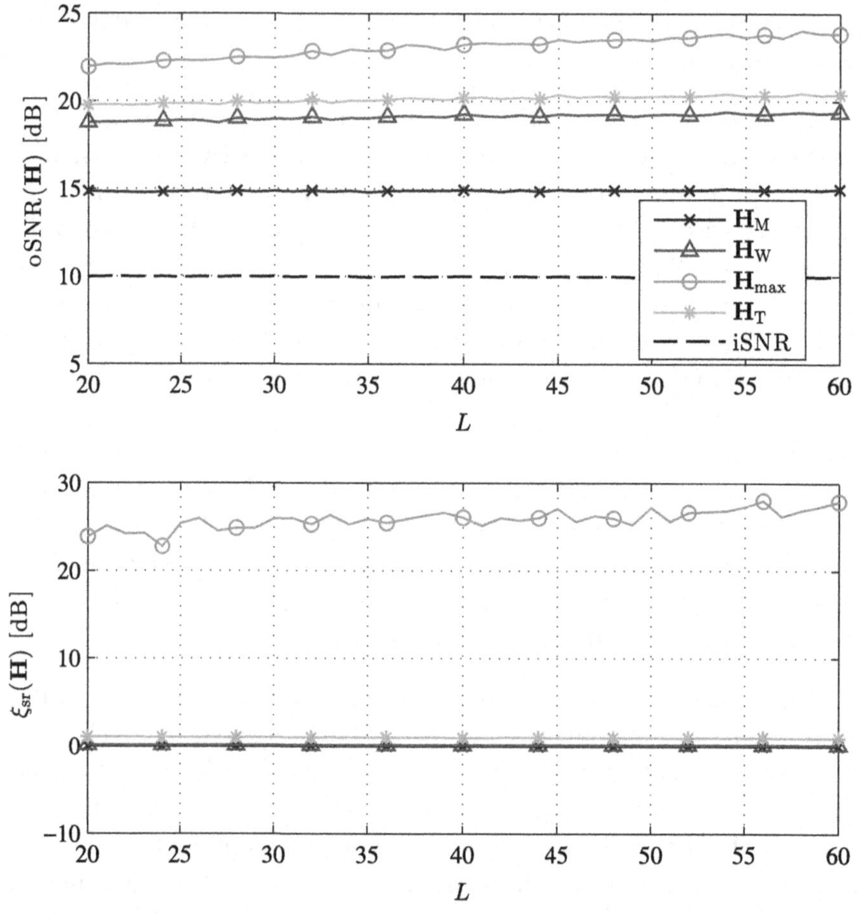

Fig. 8.12 Performance of the multichannel filters presented in Section 5.4 versus the filter length with white Gaussian noise.

distortion as it has a signal reduction factor of one. The performances were also evaluated in terms of the number of microphones as depicted in Fig. 8.13. The relationship between the filters in terms of both output SNRs and signal reduction factors is similar to the previous evaluation. The output SNRs of the filters increase at the highest rate for small M's and flatten out for $M > 6$. In the third evaluation in this signal and noise scenario, the filters' performance was considered for different input SNRs as seen in Fig. 8.14. We see that all filters have an increasing output SNR for increasing input SNR, however, their SNR gain compared to the input

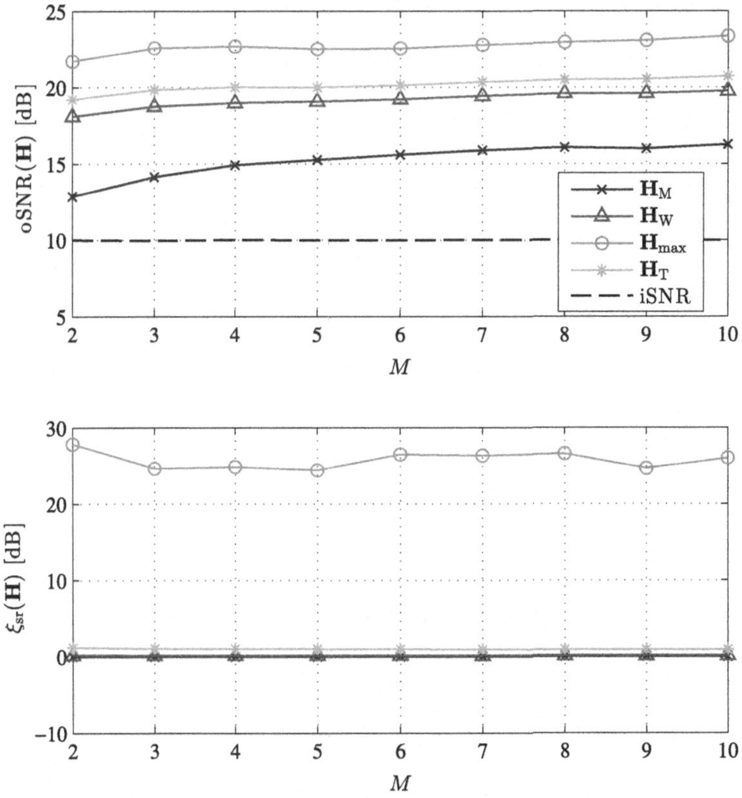

Fig. 8.13 Performance of the multichannel filters presented in Section 5.4 versus the number of microphones with white Gaussian noise.

SNR is decreasing. The signal reduction factor decreases for an increasing input SNR for all filters except the maximum SNR filter. Finally, we investigated the effect of altering the tradeoff tuning parameter as shown in Fig. 8.15. As expected, the tradeoff between the output SNR versus the signal reduction factor for the tradeoff filter can be controlled by μ. For $\mu = 0$ and $\mu = 1$, the tradeoff filter resembles then MVDR and the Wiener filters, respectively.

To also evaluate the performance of the presented multichannel LCMV filter, we considered another signal and noise scenario, where an interfering, periodic source was present in the simulated room. Moreover, white Gaussian noise was added to the signals recorded by each microphone.

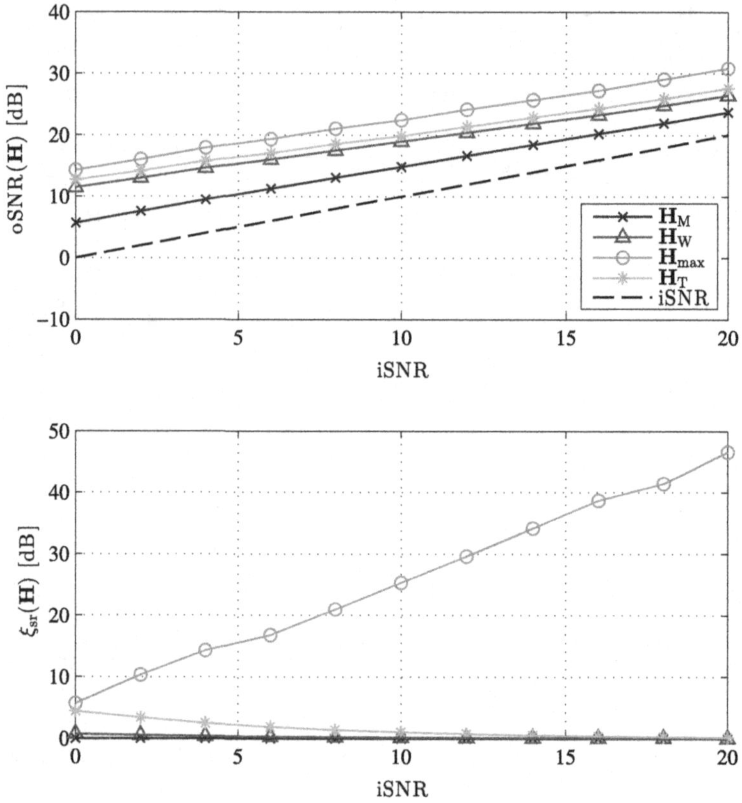

Fig. 8.14 Performance of the multichannel filters presented in Section 5.4 versus the input SNR at the first microphone with white Gaussian noise.

The periodic interferer was added at an input SIR of 0 dB at the first microphone of the array. In these experiments, the desired signal was generated in the same way as in the previous signal and noise scenario. The interfering source was assumed to be a single, real sinusoid located on the same circle around the desired source as the center of the array. More specifically, the interferer was located at an angle of $\theta_a - 90°$ on the circle, where θ_a is the angle of the center of the array. The frequency of the interferer was $\omega_i = \omega_0 + 0.2$, where ω_0 is fundamental frequency of the periodic signal used for generating the AR(20) parameters. If not specified differently, the ratio between the powers of the desired signal and the white noise (here denoted as input SNR) was 10 dB, the filter length was $L = 30$, and the number of microphones in the array was $M = 4$.

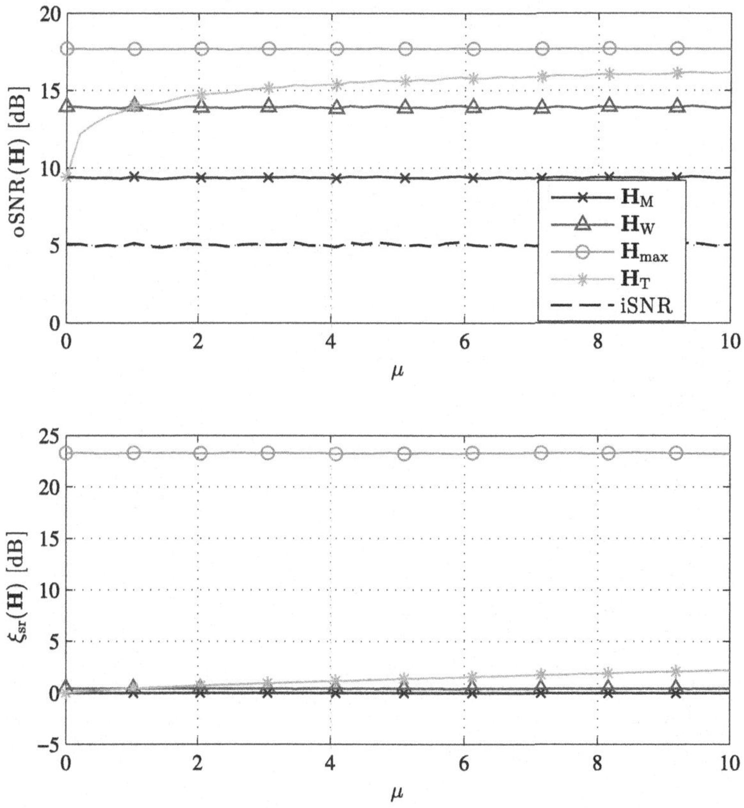

Fig. 8.15 Performance of the multichannel filters presented in Section 5.4 versus the tradeoff parameter with white Gaussian noise.

With this setup, we then evaluated all the presented multichannel filters. The first evaluation of their performances was conducted versus different filter lengths as depicted in Fig. 8.16. As expected, we make the three observations that 1) the maximum SNR filter outperforms all other filter in terms of output SNR followed by the tradeoff, Wiener, MVDR, and LCMV filters, 2) the LCMV and MVDR filters are distortionless, while the maximum SNR filter has a large signal reduction factor, and 3) the LCMV filter clearly outperforms all the other filters in terms of output SIR. In the following simulation, the filters were then evaluated for different number, M, of sensors in the microphone array, with the results being shown in Fig. 8.17. The observations from these plot are similar to those of the evaluation versus the filter length. Another

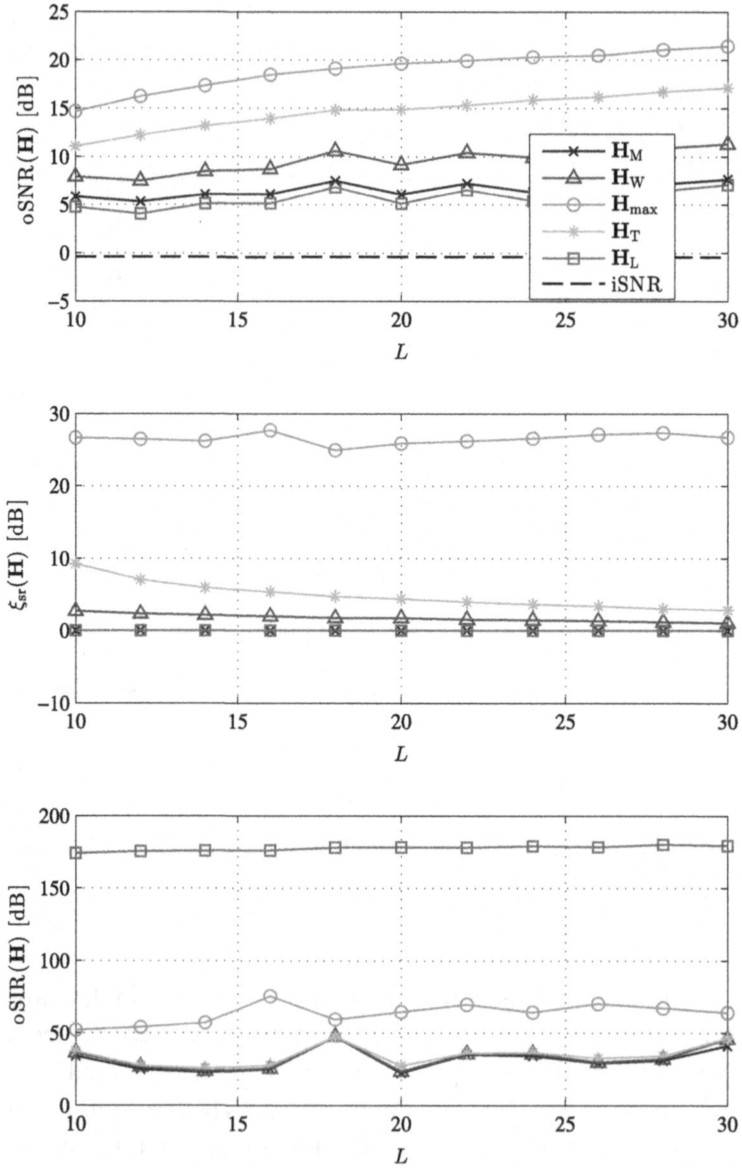

Fig. 8.16 Performance of the multichannel filters presented in Section 5.4 versus the filter length with white Gaussian noise and an interfering, periodic source.

different observation, though, is that the performance of the filters, especially in terms of output SNR, saturates after $M > 6$. Finally, as depicted in Fig. 8.18, we evaluated the filters for different input SNRs. Again, the

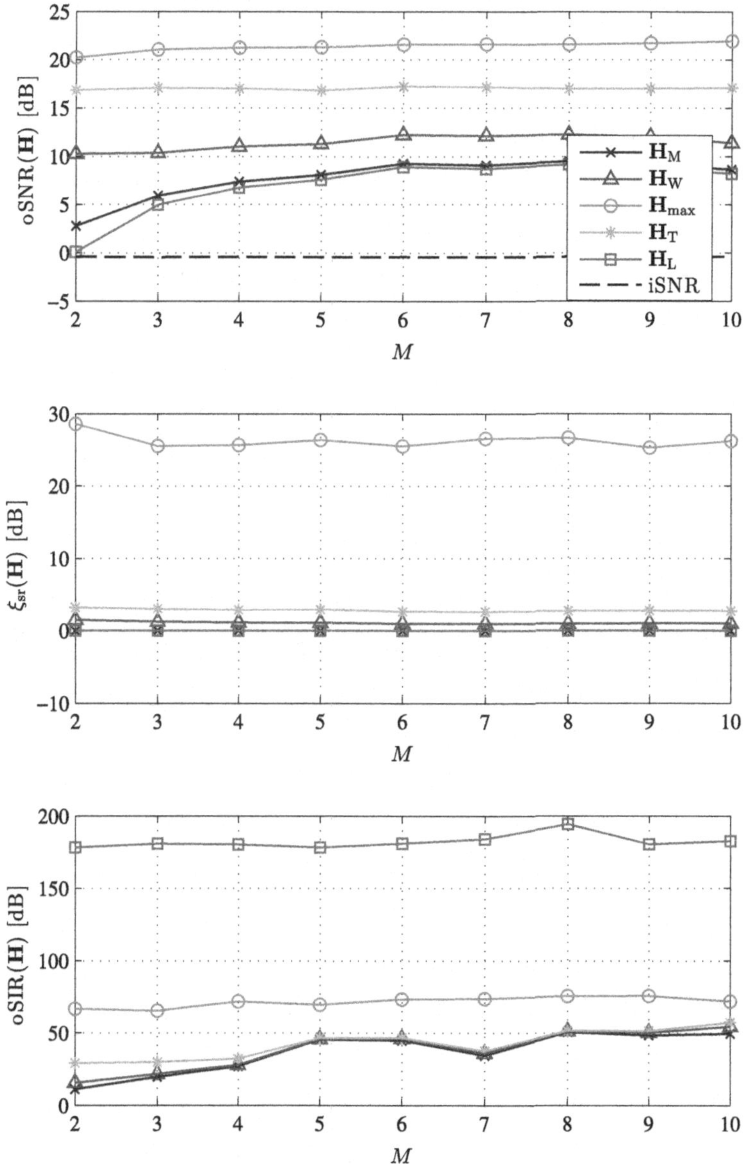

Fig. 8.17 Performance of the multichannel filters presented in Section 5.4 versus the number of microphones with white Gaussian noise and an interfering, periodic source.

filters are ranked as maximum SNR, tradeoff, Wiener, MVDR, and LCMV in terms of output SNR. The output SNR of the maximum SNR filter is increasing the most for increasing input SNR, and the output

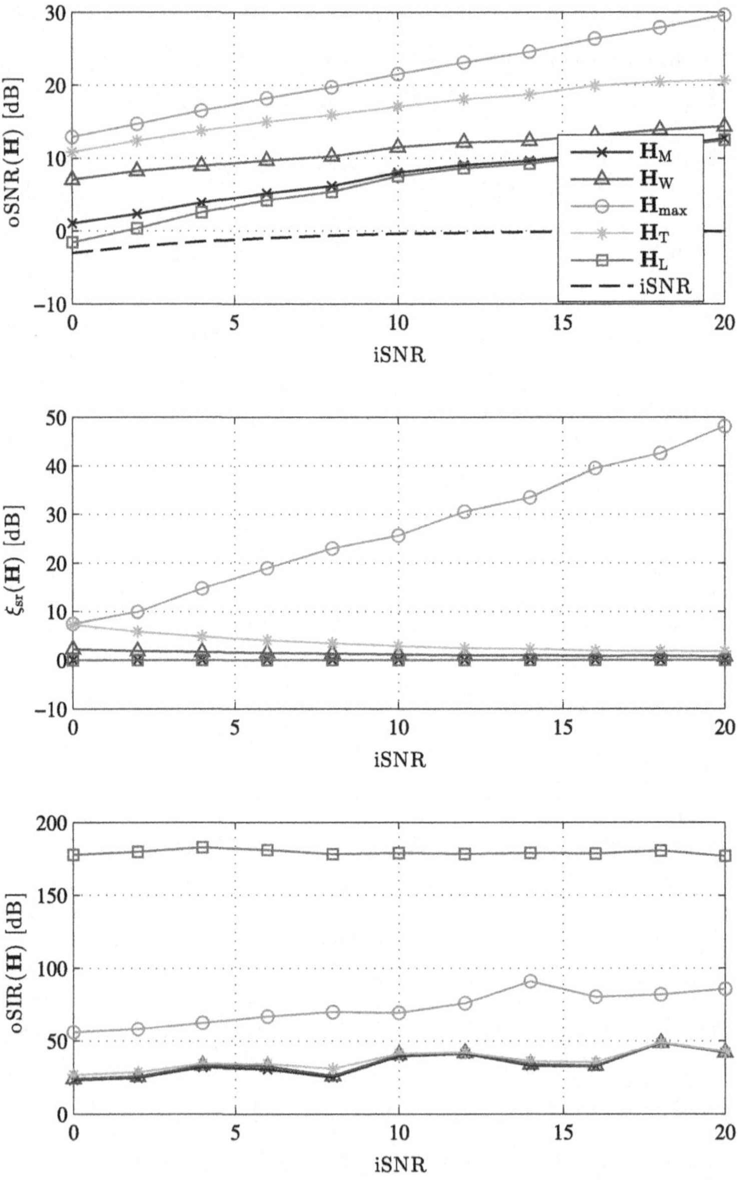

Fig. 8.18 Performance of the multichannel filters presented in Section 5.4 versus the input SNR at the first microphone with white Gaussian noise and an interfering, periodic source.

SNRs of the MVDR and LCMV filters are similar for high input SNRs (i.e., > 15 dB). We also see that the distortion, or signal reduction factor, of the maximum SNR filter increases for increasing input SNR, as opposed to the distortion of the other filters. As expected, we also observe that the LCMV filter outperforms all the other filters in terms of output SIR, with an output SIR approaching infinity.

REFERENCES

[1] P. Stoica, R. Moses, Spectral Analysis of Signals, Pearson Education, Inc., 2005.

[2] K.S. Shanmugan, A.M. Breipohl, Random Signals: Detection, Estimation, and Data Analysis, Wiley, 1988.

[3] E.A.P. Habets, Room Impulse Response Generator, Tech. Rep., Technische Universiteit Eindhoven, 2010, Ver. 2.0.20100920.

Index

Printed in the United States
By Bookmasters